MARINE CORPS TANK BATTLES IN THE MIDDLE EAST

Marine Corps Tank Battles
in the
Middle East

OSCAR E. GILBERT

CASEMATE
Philadelphia & Oxford

Published in the United States of America and Great Britain in 2015 by
CASEMATE PUBLISHERS
908 Darby Road, Havertown, PA 19083
and
10 Hythe Bridge Street, Oxford, OX1 2EW

Copyright 2015 © Oscar E. Gilbert

ISBN 978-1-61200-267-5
Digital Edition: ISBN 978-1-61200-268-2

Cataloging-in-publication data is available from the Library of Congress and
the British Library.

10 9 8 7 6 5 4 3 2 1

Printed and bound in the United States of America.

For a complete list of Casemate titles please contact:

CASEMATE PUBLISHERS (US)
Telephone (610) 853-9131, Fax (610) 853-9146
E-mail: casemate@casematepublishing.com

CASEMATE PUBLISHERS (UK)
Telephone (01865) 241249, Fax (01865) 794449
E-mail: casemate-uk@casematepublishing.co.uk

Contents

For Oscar E. Gilbert IV and Levi D. Gilbert

Preface

FOR MOST OF HUMAN history wars have been poorly documented, their stories passed down as oral histories or recorded in the self-serving memoirs of leaders. Only after World War II did the recollections and memoirs of those in the ranks became commonplace, and historians could delve more deeply into the documentation that resulted from burgeoning bureaucracies and the proliferation of typewriters. But secrecy—sometimes warranted, often not—still cloaked much. The roles of the Native American code talkers and the successes of the code-breakers that allowed the Allies to read Axis communications remained secret for decades. On the whole societal norms, official censorship, and self-censorship by the press contrived to limit public view of the worst of the blunders and horrors. The disaster of the D-Day rehearsal at Slapton Sands off the English coast "never happened." The botched "indirect approach" strategy of the 1943 New Georgia campaign (with its "alarmingly high" incidence of psychiatric casualties) was just too embarrassing.

The war in Korea was largely just ignored. It was called the Forgotten War for a reason.

Vietnam marked a seismic shift in our perception of war. Tens of thousands of highly-educated men and women were inducted (often unwillingly) at the same time that press censorship disappeared almost entirely. The result was an "in your face" war where carnage and death were splashed across the television screen at the dinner hour. Military leadership fell into disrepute, and the good suffered alongside the incompetent.

In the more recent Gulf Wars and the so-called "war on terrorism" (2001 to present) the ground shifted again. Modern wars are at once recorded in minute detail, yet poorly understood. Twenty-four-hour news coverage bombards the reader-viewer with flashing images and mind-numbing minutiae, but provides little real understanding. Frequently the necessary cloak of operational secrecy convolved with the media's compulsion to get the story on the air *now* results in a superficial understanding—or complete misunderstanding—of events that becomes embedded in millions of minds as history.

The professional military analyst (I am not one) is now drowned in a flood of data from computers, recorded message traffic, mapping and locator programs, paper records, and even old-fashioned interviews that can take years to sort through. Much has yet to be declassified, and it will again take decades before these wars are properly and dispassionately evaluated.

A new innovation—social media—resulted in another flood of more personal information: e-mails, websites, blogs, and postings. Much of it is ephemeral, and will simply disappear like the burned letters and diaries of another era.

Yet some things have remained fundamentally unchanged. Modern misconceptions to the contrary, veterans of World War II and Korea were underappreciated in their own time. They returned amid economic booms when everyone (veterans included) was experiencing an economic boom and an unprecedented rise in the standard of living. Vietnam veterans disappeared because the war was unpopular, and the unpopularity unfairly rubbed off on those who fought there.

Like Vietnam, public attitudes toward the wars of the late twentieth and early twenty-first centuries are influenced by social changes beyond the wars.

The more recent wars are different in some ways, but many fundamentals have not changed. The military in general is held in higher esteem, and even the most judgmental anti-war activists will balk at judging the individual warrior with the war. Our military is far more professional, but that professionalism has come at a cost. A sort of invisibility.

Our modern wars are fought by a small and chronically overextended minority of the population. It's the simple arithmetic of America's longest war. On the whole the seven years of the Vietnam war were fought by a succession of men serving a single twelve- or thirteen-month tour each. America's longest war has been fought by men and women who have done multiple tours; four or more are not uncommon. And they volunteered for it, so we think we need not concern ourselves overly much.

At the same time the entertainment media—particularly television and video games, but to a great degree films—have desensitized the public to the violence. The news media do a better job, but the current "may be disturbing to some viewers" warnings and presentation of videos of cold "precision" drone strikes is its own form of sanitization. Far more influential are the entertainment media.

If previous generations of the entertainment media sanitized violence and war, the current generation revels in it. Wanton violence that would have nauseated older generations of movie-goers and television watchers (just ask your grandmother) is all too often used to divert schoolchildren while mom prepares dinner. A few presentations, particularly films like *Blackhawk Down, The Hurt Locker,* and *Lone Survivor* provide a realistic and often highly emotional *approximation* of the violence of war. But television and video games just wallow in the firepower and gore for its own sake.

The result is an odd confluence of attitudes and circumstances. Our wars are fought by a better-respected but shrinking minority, and the population as a whole *thinks* it understands the experiences of that minority. But such attitudes are illusory. Seeing combat on the screen is far different from experiencing it, with its sensory and emotional overload, simultaneous physical numbing and sensory heightening, and above all the imprinted memories. So we feel free to slap a yellow-ribbon magnet on the car, think "Oh, I've played that scenario," mouth an obligatory "Thank you for your service," and go back to our games and social media.

Not much has really changed.

Acknowledgments

T HIS REPRESENTS THE final volume in a four book series on the history of the Marine Corps tank units. The primary people without whom this final volume could never have been completed are the interviewees; their names are listed in the "where are they now?" section at the end of the book. These men not only agreed to interviews, but in many cases suggested additional contacts. They represent a miniscule fraction of the many hundreds who served in Iraq, Afghanistan and other, more obscure, places.

There have been a few patient people who have been "in for the duration" of this fifteen year project. Master Gunnery Sergeant Don Gagnon (USMC, ret.) was instrumental in my efforts from the very beginning. Colonel Ed Bale Jr. (USMC, ret.) not only figured as a character in two volumes of the series, but still provides the benefit of his knowledge of the story of tanks in the Corps through our frequent conversations. The contributions of Lt. Colonel Ken Estes (USMC, ret.) have grown considerably over the years. For this volume he provided many critical documents, archival photos, and introductions to key individuals.

Sue Dillon and Greg Cina of the Grey Research Center at MCB Quantico helped track down recorded interviews and other resources. The Defense Imagery branch of the Department of Defense provided photographs, and Bill Hayes graciously provided me with his extensive collection of privately-taken photos.

The staff of Casemate Publishers, particularly Steve Smith and David Farnsworth, not only encouraged the writing and completed the publication efforts, but waited patiently while I dithered, unwilling to write the story of an unfinished war.

Of course the primary sufferers have been family. My children have provided support both emotional and technical. My wife Catherine for forty-three years has tolerated my strange work habits, and my frequent absences on business of one sort or another, has edited books, and sometimes followed me to some very strange places.

Glossary

AAV—Short for AAVP-7A1, Assault Amphibious Vehicle, Personnel, the basic version of the standard Marine amphibious tractor, often simply called an amtrac. AAVC-7A1: the communications/command version of the AAV. AAVE-7A1: the unofficial designation for the engineer vehicle equipped with a roof-mounted Mk154 MICLIC launcher.

BLT—Battalion Landing Team, a Marine infantry battalion and supporting elements.

CAX—Combined Arms eXercise.

COIN—COunter INsurgency

Company Wedge—A line of tanks in an offensive "arrowhead" with one platoon in front, two others staggered back and to the flanks.

CSSC—Combat Service Support Company, provides engineer and logistical services

Defensive Coil—At the halt the tanks face outward in a 360degree defensive circle, with thick frontal armor outward and soft-skinned vehicles in the center.

DPICM—Dual Purpose Improved Conventional munition is an artillery round that scatters small contact-detonated bomblets to attack infantry and vehicles. the artillery version of a cluster bomb.

Dragon Eye—The smallest Unmanned Aerial Vehicle in US service. Launched by an elastic cord and powered by a lithium battery, it can be fitted with a variety of sensors. It can be flown from the ground or preprogrammed. Upon completion of its one-hour mission the controller crashes it into the ground, where the major components break apart for recovery and quick reassembly.

FCP—Forward Command Post.

Fedayeen—"Those who sacrifice," a quasi-militia organized by the Ba'athist Party in Iraq.

FSSG—Force Service Support Group, responsible for logistical support.

Gypsy rack—The exterior rack on the back of a tank turret used to carry miscellaneous gear.

HEAT—High-Explosive Anti-Tank.

HEMTT—Heavy Expanded Mobility Tactical Truck, any of several types of ten-ton 8x8 trucks.

IFAV—Interim Fast Attack Vehicle, at this period a "military dune buggy".

Javelin—the FGm-148 "fire-and-forget" self-guided shoulder-fired anti-tank missile.

Jihad—roughly "struggle," the term has entered Western vernacular as "holy war," but the word is far more complex, including any struggle of faith, commonly the struggle to live a sanctified life, against personal moral weakness, and against oppression or social injustice.

LAI—Light Armored Infantry. The LAV-equipped battalions were originally conceived as mechanized infantry, but they were too lightly-equipped for this role.

LAR—Light Armored Reconnaissance. The role of the LAV-equipped battalions was re-evaluated, and their role was changed to mechanized reconnaissance and force screening.

LAV—Light Armored Vehicle, the eight wheeled armored vehicle used by the Marines. LAV-25: the basic vehicle armed with a 25mm chain gun, carrying four infantry dismounts. LAV-AD: the Air Defense variant armed with a chain-gun and eight Stinger missiles. LAV-AT: the Anti-Tank variant armed with a two-tube TOW missile launcher. LAV-L: the cargo carrier version; often use as an armored ambulance. LAV-M: 81mm mortar carrier.

LCAC—Landing Craft Air Cushion, a large hovercraft capable of carrying a tank.

LPD—Landing Platform, Dock. Any of several classes on naval vessels that can launch either landing craft or AAVs carried inside and helicopters from a flight deck. An LPD has a hangar for servicing aircraft.

LSD—Landing Ship, Dock. In current usage similar to the LPD, but without an aircraft hangar.

LTI—Limited Technical Inspection—a checklist for preventive maintenance.

MEF—Marine Expeditionary Force, a Marine Division and attached air and support assets.

MEU(SOC)—Marine Expeditionary Unit (Special Operations Capable). A force built around an infantry battalion with attached supporting units such as artillery, armor, air assets and logistical support. the SOC suffix indicates a high level of training and the capability to undertake "special operations" missions.

MLRS—The m270 multiple Launch Rocket System, with reloadable launcher pods mounted on a tracked chassis.

MOS—Military Occupational Specialty, the Marines' job description.

MOUT—Military Operations in Urban Terrain.

MPS or **MPPS**—Maritime Pre-Positioning Ship/Squadron, refers either to individual ships or to squadrons of cargo ships loaded with heavy weapons and supplies, and stationed at strategic points around the globe. Marines can be airlifted to "marry up" with the weapons and supplies, greatly speeding deployment to trouble spots.

MSPF—Maritime Special Purpose Force, specialized sub-units within a MEUSOC undertaking special missions such as covert reconnaissance or hostage rescue.

MSR—Main Supply Route.

MTVR—Military Tactical Vehicle Replacement, a 7-ton 6x6 cargo truck built by Oshkosh for the Marine Corps and Navy. The design provides unusually good off-road capability.

MPAT—Multi-Purpose Anti-tank. this multi-purpose HeAt round can sense the nature of the target, acting as a HEAT round to penetrate armor or as an He round to penetrate and explode inside soft targets.

MULE—Modular Universal Laser equipment, a big, box-like target designator usually mounted in front of the loader's hatch on a tank.

Phase Line—A line on a map, usually some easily identifiable terrain feature use to align units to avoid friendly fire incidents. The practice dates to World War II.

Reactive Armor—Sometimes called Blazer armor though not an official term. Blocks of explosive in thin metal shells spaced a few inches from the tank's main armor, their impact detonation will disrupt the penetrating jet formed by HeAt rounds. It provides no protection against APDS (sabot) rounds.

REDCON—REaDiness CONdition.

REIN—A suffix used to indicate a unit is reinforced by assets from another unit.

REMF—REar area Mother-Fucker, coined in Vietnam to indicate anyone not in a front-line unit.

Sagger—Originally the NATO code name for the older Soviet-made AT-3 anti-tank missile, it was used generically for any enemy anti-tank missile.

SIGINT—SIGnals INtelligence

SINGCARS—A secure radio that "hops" frequencies according to a pre-programmed, time-keyed sequence.

Short-Tracked—"Short-tracking" compensates for a missing wheel or broken suspension arm on the first or last wheel position. the wheel is removed, the suspension arm is removed or wired into the up position, and several track blocks removed to make the track shorter.

SOTG—Special Operations Training Group, provides specialized pre-deployment training for MEUSOCs.

Stinger—The US-made FIM-92 shoulder-launched infrared homing anti-aircraft missile, it also equipped the LAV-AD.

TAC—TACtical command post.

TOW—The BGm-71 tube-launched Optically-tracked Wire-guided first-generation heavy anti-tank missile.

Tank Tables—Prescribed fire and mobility exercises involving pop-up targets to be identified and engaged with various weapons. they are arranged in order of increasing difficulty, one through twelve. Combat is sometimes called "table thirteen."

Al-Wahhabi—literally "the generous donor," perhaps an oblique reference to wealthy Saudi backers. Adherents prefer *al-Muwahhiddun* ("the monotheists").

VTR—Vehicle Tank Recovery, a heavy tracked vehicle used to recover disabled tanks, at this period the M88 family.

ZSU-23-4—The *Shilka* is a radar-controlled four-barreled 23mm antiaircraft gun on a tracked chassis. They—like all anti-aircraft guns—were priority targets because of the threat to low-altitude aircraft.

Prologue

Victorious warriors win first and then go to war, while defeated warriors go to war first and then seek to win.—Sun Tzu

Battles are won by slaughter and maneuver. The greater the general, the more he contributes in maneuver, the less he demands in slaughter.—Winston Churchill

Most battles are won before the first shot is fired. Some are won years earlier, around a polished wooden table.

In the aftermath of the Vietnam War junior officers and NCOs who would one day rise to high positions were determined to change the culture of the Marine Corps, and indeed the entire US military. This new generation had seen other junior officers and young enlisted men suffer and die in a numbing sequence of battles fought for no clear goal. They realized that the American public would no longer tolerate the costly frontal assaults that the Marines had of necessity endured in the island-hopping campaigns of World War II, the trench warfare for useless rock piles like those along the Jamestown Line in Korea, or least of all the endless and ultimately futile battles of attrition of the Vietnam conflict.

A new generation of Marine Corps leaders would emphasize maneuver over slaughter, paralyzing an enemy and his will to fight by rapid maneuver combined with the limited and precisely directed application of overwhelming violence. Interestingly the tank, the very symbol of mindless mechanical destruction, would play a key role in this new scheme of limited, more "humane" war. The tank's cannon and machine guns could apply deadly force far more precisely, in a way that minimized "collateral damage," the inadvertent death and destruction wreaked by air strikes and artillery fire, the

17

blunt instruments that America had previously relied upon in its wars.

Martin R. "Marty" Steele had served as both an enlisted man and junior lieutenant in Vietnam, and later as a company commander in both tanks and amtracs. He was the first Marine to be the Distinguished Graduate of the joint Army-Marine Corps tank school at Fort Knox. As a field-grade officer he stood up the 1st Light Armored Vehicle (LAV) Battalion, commanded the 1st Tank Battalion, and contended with Chrysler Corporation executives to square away the troubled M60A1 tank procurement program. But as a passionate tank advocate his most crucial battle would be against the Marine Corps' institutional inertia.[1]

Five Commandants ultimately wrestled with the issue of modernizing the tank fleet, and Steele was a thorn in the flesh of each. In 1981 the Corps had decided, *in principle,* to upgrade its tank fleet by adopting the new M1 series, the tank the Army called the Abrams. Then they declined to implement the decision. In the eternal competition for a share of the budgetary pie, many senior officers advocated for the "Super M60" program pushed by the tank's manufacturer. Glowing reports by company executives who wanted to keep the M60 production program alive lent weight to the arguments of those who wanted to spend the limited funds on other programs— aircraft, improved Assault Amphibious Vehicles (AAVs), new LAVs, and a host of other weapons. New tanks kept slipping down the list of priorities.

In 1987 General Al Gray, one of the most outspoken advocates of maneuver warfare, became Commandant, but the Corps still did not have the modern tanks that would be essential to implement the new doctrine. By 1988 Steele was still on what seemed like a suicide mission, still advocating for the long-delayed replacement of the aging fleet of M60A1s. Steele's morale was at its lowest ebb when one bitterly cold, gloomy Sunday morning in February he went for a walk in Arlington National Cemetery to clear his mind. On his icy walk it began to rain and sleet, covering everything with a sheet of ice. Nearly blinded, Steele literally stumbled over a tombstone; the grave of General Creighton Abrams, the namesake of the new tank and with whom he shared a birthday, 15 September. Even for a religious man, it was as close to divine endorsement of a mission as he was likely to be granted. The next day Gray approved the M1 acquisition. In principle.

In 1990 Steele still struggled to implement the M1A1 acquisition program, and on the eve of war finally gained the support of Maj. Gen. Jack

Sheehan, one of Commandant Gray's closest confidantes. He convinced Sheehan that despite the assurances of many, the M60A1 was a dying creature. The Corps simply did not procure enough tanks for a stand-alone program, and the Army would no longer support a system for spare parts and supplies like 105mm ammunition to keep the obsolescent M60A1 viable. When Sheehan recommended approval, Gray finally issued directives for Reserve and Regular units to acquire the new tank, but on a limited basis. Like the M26 in the late 1940s, the new tank would appear only in a few test units.

It seemed that the Corps' tankers were finally a bit closer to having a vehicle so markedly superior to its potential foes that it bestrode the battlefield like some nearly invincible seventy-ton colossus, able to destroy precisely at long range in rain and darkness while remaining far less vulnerable to the enemy's blows. Yet key aspects of tank warfare remained unchanged.

In order to better detect threats to his vehicle and crew, and to execute his mission of supporting the infantry, the tank commander's greatest asset would remain the unarmored eyeball. To better see the battlefield—maintaining what the military calls "situational awareness"—and direct his weapons, the tank commander would risk his own life to ride with his body partially exposed high atop a "bullet magnet."

As in all wars past, the greatest threat to the tank and its crew would come not from enemy tanks or guns, but the stealthy mine in its new incarnation, the command-detonated improvised explosive device, or IED.

The tank would also finally have an able stable-mate to share some of its more odious traditional missions like convoy escort and mechanized reconnaissance in high-risk environments. In its constant quest for "lightness"—strategic mobility—the Corps had experimented with fast, lightly-armored wheeled vehicles like the King armored car (1915–1934), the White M3A1 Scout Car (late 1930s–early 1940s), and the M8 armored car (limited use in Korea). None had proven particularly successful, and everyone was painfully aware that they were not tanks. The new wheeled Light Armored Vehicle, or LAV, would for the first time provide the combination of firepower, armored protection, and mobility (though as with all light armored vehicles, limited in all three) plus transportability that the Corps constantly sought.

Both the new tank and the LAV would soon be tested in a new combat arena, the deserts and mountains of the Middle East and southern Asia.

NOTES

1 The account of Steele's travails is condensed from Estes, *Marines Under Armor*, p. 183–185, and Jones, *Boys of '67*, p. 313–323.

CHAPTER ONE

A Brutally Complex World

*God does not change what is in a people, until they change
what is in themselves.*—The Koran, Chapter 13, verse 11

THIS IS NOT A story of religious wars. Rather it is the story
of one country spending treasure and the lives of its citizens to hold together the remaining bits of broken empires.
It is fashionable to see the troubles of the Middle East and beyond solely
as a struggle between the forces of militant Islam and more secular Western
cultures. Make no mistake, humans fight savagely over differences of religion. Religious minorities were and are persecuted. However, far more of
the conflict that has plagued the Middle East stems from mundane but familiar evils: blind nationalism, greed, and naked ambition.

Much "religious" conflict stems simply from the very human tendency
to use religion to justify sordid personal ambitions. We can clearly perceive
human faces behind the religious masks of the nationalists of Revolutionary
Iran as it backs Hezbollah (Party of God), or the secular tyrant Saddam
Hussein's cynical invocation of Qadissayah (the epic battle that determined
the path of mainstream Islamic dynastic succession and still divides Sunni
from Shia) as a nationalistic rallying cry.

For most of documented Middle Eastern history, religious persecution—when it flared—was largely a local matter. Because of difficulty of
travel and communication the flames of any *farhud* (religious purge) would
burn themselves out before reaching the scale of a national or international
disaster. In contrast, secular empires, and later national governments, had
the means and the abilities to impose long-lasting and thorough persecution.

Iraq, around which most of this story inevitably centers, has had the misfortune to sit at the intersection of many secular empires as well as lie in the path of barbarian invaders from the steppes of Asia for thousands of years. In the modern era since 1918, both Iraq and the surrounding modern states have been a hotbed of intrigues, dynastic struggles, coups, and wars. Even an incomplete history of Iraq alone is beyond the scope of this book.[1]

After the fall of the decadent Abbasid Caliphate (AD 1258), and rule by the Mongols, Tamerlane, and Persia (modern Iran), in the early sixteenth century, Iraq—like most of the Middle East—fell under the rule of the Ottoman Turks. The Ottomans were astute rulers, adopting many of the practices of the Byzantine Empire, successors of the eastern Roman Empire, which they overran in 1453.

They provided for more or less local governance by subdivision into *vilayets*, or provinces, along ethnic and tribal boundaries. Under the Ottomans modern day Iraq was divided among three *vilayets*, governed from Mosul (Kurdish), Baghdad (Sunni), and Basra (Shia). Within *vilayets*, local rule was exercised through family or tribal chieftains acceptable to the locals. The proximity of the powerful Persian Empire helped assure the protection of the Shia minority.

The Ottoman ruler assumed the title of Caliph, but for most of the Empire's life the real power lay with the *janissaries* and the *mameluke*. In a legalistic nicety to sidestep the Koran's proscription against Muslim warring upon Muslim, the Ottomans created these classes of warrior slaves from Christian Europe and the Asian steppes, respectively. Eventually those who wielded the weapons wielded the power, and they governed semi-autonomously, the *janissaries* from Constantinople, the *mameluke* from Baghdad and Cairo.

The Ottoman military castes were frozen in time and tradition, and between 1826 and 1834 the power of both castes collapsed. As the power of the Caliph also eroded, secular nationalists came to power, culminating in the rise of nationalist factions within the Turkish Army.

Along the fringes of the dying Empire factions struggled for control over land no major power coveted. Between 1902 and 1936 Ibn Saud and Abdul Aziz slowly consolidated the desert kingdom of Arabia. Another obscure bit of desert that was competed for by two empires lay at the extreme western end of the Persian Gulf. Part of the *vilayet* of Basra, this patch of desert had had few natural resources, but controlled waterborne commerce from the immense Mesopotamian drainage basin, which included the Ottoman Empire's richest regions. Basra was a major port for the *haj*—the Islamic pilgrimage to Mecca, and was a producer of pearls. Under the protection of the Royal Navy and the administration of India, this strategic bit of sand, arbitrarily proclaimed the Emirate of Kuwait, became a British protectorate in 1897.

A seismic shift came with the ill-fated Ottoman decision to side with Germany in the Great War. In October 1918 the Ottomans agreed to an armistice with Britain, and their Empire was partitioned into several protectorates that would reshape regional history.

The French assumed mandate responsibility for modern Syria and Lebanon; the British took Palestine, Transjordan and Iraq. The British favored the Hashemite dynasty as hereditary rulers. The Hashem, descendants of The Prophet Mohammed's clan, should have been naturally acceptable, but too many other factors, not least the first stirrings of nationalism and old-fashioned ambition, were at work. Only Abdullah I of Jordan established an enduring Hashemite dynasty. Ruling until his assassination in 1957, his family succession continues today.

In Iraq the British ignored major regional and ethnic partitions and promoted the Sunni minority which dominated only the deserts of the west. They were immediately faced with revolt, and by 1920 fighting had spread throughout the country. The revolts cost the lives of 6,000 Iraqi and 500 British and Indian soldiers.[2] The British eventually established a colonial government in which many of the senior posts were held by the more amenable Sunnis. They also established a second army, the Iraqi Levies, answerable only to the British, consisting of Kurds, Marsh Arabs,[3] and Nestorian Christian refugees from Turkey.

In neighboring Syria the republican-minded French backed the Hashemite Amir Faisal.[4] His rule lasted only a few months, and the French were soon faced with resistance culminating in a revolt which they brutally repressed in 1925. A treaty promised independence in 1936, but the French reneged on the agreement.

The French drew the boundary between Syria and Lebanon, long a semi-autonomous province under the Ottomans. The border was moved to the crest of the mountains east of the Shia- and Druze-populated Baqaa

(Bekka) Valley, a logical decision in European eyes but one which ignored long-established ethnic and tribal boundaries that had placed the valley firmly in the Damascus *vilayet*.[5]

The British established a Hashemite dynasty in Iraq, but chose the Europeanized Amir Faisal of recent Syrian nationality, who lacked both nationalist credentials and local acceptance. The cobbled-together state faced threats foreign and domestic. Turkey was loath to give up the riches of northern Iraq. The Kurdish governor declared himself king of a Kurdish state in the north; the British immediately drove him into exile in Iran.

For thousands of years men had known that natural gas vented from the ground, and Plutarch had written that the locals set an oiled road afire to impress Alexander the Great.[6] On 15 October 1927 a consortium of British, Dutch, French, and American companies discovered oil in vast quantities near Kirkuk in Kurdish Iraq, establishing the Middle East as a rival to Russia and the East Indies for oil production.

The fledgling Iraqi government council finally persuaded Britain to set a date for independence, October 1932. The new Prime Minister, Nuri al-Sa'id, was a progressive, a modernizer, and a pragmatic nationalist. But following independence Sunni landowners and army officers soon became the real powers.

The Iraqi Hashemite dynasty was long-lived, unlike its members; Faisal reigned until 1933, and was succeeded by his son Ghazi. Under Ghazi the army first began to flex its muscle in domestic politics, staging a coup in 1936 to install a cabinet more to its liking. Ghazi died in 1939, succeeded by his three year-old son Faisal. The uncle of three year-old Faisal, Abd al-Ilah, served ineffectively as regent, unable to control the army, who openly favored the rising dictators Hitler and Mussolini.

In 1938 Anglo-Persian and Gulf Oil companies discovered oil in Kuwait, followed within weeks by discoveries in Saudi Arabia. Soon British interests had to compete with a flood of representatives from oil-starved Italy, Japan, and Germany. The outbreak of war in Europe stifled further exploration, as American and British explorers returned home.

Following the fall of France to the Nazis in 1940 the Vichy government controlled Syria and Lebanon. In early 1941 Prime Minister Rashid Ali al-Kailani tried to bar British use of treaty bases in Iraq. After a coup the regent went into exile in British Transjordan. Ali sought German aid; the Vichy government allowed a small force of German planes to transit Lebanon and Syria into Iraq. A British mechanized column ("Habforce") from Transjordan drove out the Germans and restored the pro British faction.[7] After the German incursion, the British and Free French seized Syria and Lebanon. Iraq's Kurds and Shia remained restive, and for the first time Communist agitators appeared, exploiting nationalist ambitions.

An obscure supporter of Rashid Ali was Khairallah Tulfah, a member of the fledgling Ba'ath ("Arab Socialist Resurrection") Party and maternal uncle and patron of a small boy rejected by his stepfather. The boy, Saddam Hussein, had been born between 1935 and 1938. When his uncle was imprisoned following the failed German incursion Saddam's stepfather drove the boy from home, and he made his living as a very young petty criminal.

In 1943 Lebanon declared independence, and the Free French briefly imprisoned the entire government. France eventually abandoned Syria and Lebanon in 1946. For Lebanon the French-mandated disruption of ethnic partitions resulted in a weird constitution in which the President was required to be from the Maronite Christian community, the Prime Minister a Sunni Muslim, and the Christians were in practice guaranteed a six-to-five dominant ratio in the Parliament, regardless of shifting demographics. It was a recipe for ongoing disaster.

The second seismic shift in regional affairs was the foundation of the state of Israel. In 1917 the secret Balfour Declaration expressed sympathy for formation of a Zionist state. Jewish settlers began to flow into Palestine, growing to a flood in the aftermath of World War II. In May 1948 Israel proclaimed itself as an independent state at the expiration of the British Palestinian mandate. Almost immediately the remnant Arab Palestinian state, Syria, and Egypt were involved in war with the new state.

Iraq, with no common border, was not directly involved but sent troops to newly independent Jordan (formerly Transjordan), probably helping to forestall Israeli conquest of the West Bank. Jordan had an almost tacit agreement to limit fighting with Israel; Jordan simply annexed the West Bank region of Palestine. The short war drove a flood of Palestinian refugees into adjacent states.

Soviet influence continued to grow, particularly among junior military officers; in 1952 a coup overthrew the Egyptian monarchy, bringing Gamel Nasser to power. Iraqi Prime Minister Nuri al-Sa'id, in and out of power as the result of internal bickering, was of course wary of this growing Soviet regional influence. He was instrumental in forging the Baghdad Pact, a regional alliance with Turkey, later expanded to include Pakistan, Iran, and Britain.

Humiliated in its battles with Israel, in 1956 Syria invited the devil to the party and entered into a pact with the Soviet Union. In the same year Egypt nationalized the strategic Suez Canal, and gained considerable regional influence through successful confrontation with Britain and France. Soviet efforts to rule the region by proxy defined relations with the United States for decades.

In February 1958 Nasser's Egypt and Syria formed the United Arab Republic (UAR), ostensibly forming a single country. It prompted a similarly temporary Arab Union of the Hashemite monarchies, Iraq and Jordan.

BEIRUT PRELUDE, 1958

The Maronite Christian President of Lebanon, Camille Chamoun, had sided with Britain and France in the Suez Crisis of 1956, and openly favored the Arab Union, offending Arab pan-nationalists including his own Prime Minister. Domestic unrest included destruction of the US Information Agency offices. The United States viewed the situation with increasing alarm, in the mistaken belief that the UAR was a Communist front.

Fearing Syrian invasion, and with the counterbalance of Iraq at least temporarily disabled, Chamoun requested help from the US. President Eisenhower invoked the self-proclaimed Eisenhower Doctrine, stating that the US had the right to intervene in nations under threat from international Communism. The agents of the intervention would be Army troops from Europe, and two reinforced battalion landing teams (1/8 and 2/2) of Marines.

Mounted on twelve hours notice, the Marine landings on 15 July were a bizarre spectacle. Beirut was a major tourist destination, and the landing force was under strict orders not to interfere with beachgoers. The Marines of 2/2 stormed ashore near the airport amid swarms of bikini-clad tourists (it was still that kind of country) and local villagers. Teenagers helped the Marines drag heavy gear through the surf, while other civilians cheered. The force first secured the airport for the arrival of Army airborne units from Germany.

Early the next day tanks come ashore across a floating causeway built by the Shore Party. The commander of the 2nd Provisional Marine Force, Brig. Gen. Sidney Wade, led a column of vehicles including the only two tanks available, into the city to secure the port, critical bridges, and the embassy complex.

There was reason to be nervous. General Wade had gone into the city to meet with the US ambassador and General Fuad Shihab.[8] Shihab was concerned that his troops might fire on the Marines. Then word arrived that Lebanese Army tanks had established a blocking position along the highway. Wade drove toward the airport in an embassy car and encountered the Lebanese tanks; a short conversation revealed that they had no clear orders, but would probably resist any American advance.

At 1100 hours the Marines, still under their original orders, mounted up and moved toward the city.

PFC John E. Dreisbach was a radio operator.

"On the long column—containing tanks, amtracks, trucks and jeeps—that entered Beirut, I was sitting on a jeep fender," remembers Dreisbach. "There were thousands of people lining the streets, half of whom seemed to welcome us. The looks of the others made me get off the fender and jump into a truck, as they didn't look that friendly. I was also uneasy about the buildings, which seemed to be damaged by explosives."[9]

About a mile north of the airport, the column encountered the Lebanese tanks, whose guns were trained on the Marines. The Marine rifle battalion commander had halted the advance when cars racing to and fro—Wade and naval officers headed into the city, the ambassador and General Shihab headed out—arrived at the roadblock.

In a hasty conference in a nearby schoolhouse Shihab and the Americans worked out a plan. Each Marine unit would be preceded by Lebanese Army representatives, and others would be "embedded" (a term not then in use). The Marines would be careful to bypass the Muslim quarter of the city.[10] The truce was fortunate, since the Marine tanks had no main gun ammunition; the ship carrying it and the balance of the tank platoon had gone to Malta for repairs.[11] By 1900 hours the Marines had secured their objectives.

When 1/8, with fourteen more tanks as well as LVTs and Ontos, arrived the Marines were able to undertake mechanized patrols. These few vehicles (fifteen tanks, ten Ontos, and thirty-one LVTP-5s) represented the total US mechanized presence until the arrival of Army tanks on 27 July.[12]

Ensuing confrontations were curiously polite by the standards of another time. Marines taken prisoner by anti-government Islamic forces were lectured at length and then released, and their weapons returned. Marines staged practice landings, to the amusement of beachgoers.

The last Americans withdrew on 25 October; the occupation resolved the crisis when Chamoun was persuaded to resign, replaced by the more pragmatic General Shihab. The cost to the US for this successful operation was unbelievably minimal: one soldier killed and one wounded.

Following a 1958 coup, power in Iraq was consolidated in the person of Abd al-Karem Kassem, an Army general of mixed Sunni and Kurdish Shi'a ancestry. Another reformer, he fell to grief when he allowed the Kurdish Democratic Party to operate openly, only to face revolt as the Kurds demanded autonomy. This was the beginning of the Kurdish problem.

Kassem's hold on power was threatened by multiple coups, including one by the Ba'athists in March 1959 that led to bloody reprisals. Saddam Hussein was by now an obscure but ruthless junior operative in the Ba'ath Party. He had proven himself to his superiors by murdering Saddoum al-Tikriti a Communist who had denounced Saddam's uncle. Saddam Hussein attempted to assassinate Kassem, but botched the job and fled into a four-year exile.

Kassem laid the foundation for much future mischief, asserting Iraq's right to control the mouth of the strategic Shatt el-Arab waterway, asserting ownership of Kuwait, and took the first steps toward nationalizing Iraq's oil industry. Britain, along with the UAR, disabused him of the idea of annexing Kuwait. Iraq found itself increasingly isolated, and worse, humiliated by failure, in the Arab world.

The Ba'athists were by now a growing force in Iraqi politics. Kassem arrested some Ba'athists, but the measures were ineffective. On 9 February 1963 Ba'athist army officers seized control of Baghdad, and Kassem was murdered. The young Saddam Hussein returned from exile. The one thing his uncle's patronage could not provide was admission to officer's training in the Iraqi Army, still a bastion of the well-born and educated. Despite his loyalty and service to the party, he was snubbed by the officer corps, ever a power in Iraqi domestic politics. He came to admire dictators—particularly Stalin—who had risen through the power of their own will.

In November the Iraqi Ba'athists were in turn overthrown, and a nationalist, Abd al-Salim Arif, gradually consolidated power. He suppressed the Ba'ath and created the Republican Guard as a reliable personal army. Saddam was again imprisoned following a failed counter-coup in October 1964. By now Saddam had developed a network of influential patrons who thought they could harness his single-minded ruthlessness. Rather than being executed, Saddam escaped prison under suspicious circumstances.

Arif also brought to fruition the nationalization of the oil industry, forming the Iraqi National Oil Company (INOC) to develop oil reserves that European and American companies were holding in reserve. Arif was succeeded by his less astute brother, Abd al-Rahman Arif, after his death in a helicopter crash.

CARIBBEAN INTERLUDE I — THE DOMINICAN REPUBLIC, 1965

The Marines are no strangers to either Haiti or the Dominican Republic, the two nations who share the troubled island of Hispaniola.[13] One of the oldest European colonies in the hemisphere, the Dominican Republic proclaimed its independence in 1821, only to be promptly occupied by forces from neighboring Haiti until 1844. Spain then took advantage of the American Civil War to reoccupy the Republic from 1861 until 1865. After regaining its independence, the unhappy country was victim to a dizzying round of coups and revolts, and Marines were landed to suppress troubles in 1903 and 1904.

Civil wars and corruption had driven the nation deeply into debt to French and German banks, and the U. S. feared that European powers would use this as a pretext to establish bases that could dominate sea approaches to the Panama Canal. In 1907 the US took over control of the Dominican customs service, and thereby the national economy, and Marines moved in to police the country. Though the Marines strove to establish a professional national police and improved the national infrastructure, the occupation grew increasingly unpopular in America, and the Marines were withdrawn. In 1930 Rafael Leonidas Trujillo seized power, establishing a dictatorship.

On 30 May, 1961 Trujillo was assassinated. Following rule by a series of rotating juntas, the elections of early 1963 brought an exiled leftist reformer, Juan Bosch, to the presidency. His term was short-lived. A September military coup, and another series of juntas, brought Donald Reid Cabral to the Presidency, opposed by Bosch leftists as well as military factions. On 24 April 1965 dissident Army officers arrested their own Chief of Staff, and by evening the government had indeed fallen. Bosch ("Constitutionalist") supporters seized caches of weapons, and by 27 April the capital was racked by sniping, looting, and arson. One military faction, the "Loyalists," tried to fight its way into the city, only to be repulsed by the rebels. On the afternoon of 27 April the rebel president, Molina, requested American aid in restoring order.

The Johnson administration was predisposed to support the Loyalist forces. On 28 April, Marines from the USS *Boxer* landed by helicopter to begin the evacuation of foreign citizens. The following day the 6th MEU (India Company, 3/6) and supporting armor landed west of the capital. The Marines moved rapidly toward the city. Shortly after midnight on 30 April planes carrying the first Army airborne troops landed at San Isidro Airfield east of the city.

That afternoon paratroopers of the 3rd Brigade and Marines began to move into the city as Loyalist troops unexpectedly withdrew. The part of the city to be seized by the Marines of 3/6 would constitute much of the International Safety Zone (ISZ), but contained more rebel forces. Company columns, led by small task forces consisting of a rifle platoon supported by two LVTs and a tank, pushed into the city. Other Marines followed in trucks. Forbidden to use heavy weapons, the Marines were pinned down by heavy fire until permission was granted to use 3.5inch rocket launchers. On 1 May the Marines and airborne soldiers linked up, splitting and isolating the Constitutionalist forces.

Over the following week more forces poured in including the remainder of Bravo Company (-) (REIN), 2nd Tank Battalion.[14] Army forces engaged and destroyed two armored vehicles, but Marine armor faced no opposition. American forces were able to quickly pacify the nation.

Forbidden to use their heavy firepower, the Marine tanks nevertheless served a purpose that would be increasingly important in future operations of this type—intimidation.

———

For nearly a decade the attention of the American public was fixated on the war in Vietnam.[15]

The humiliation of Arab armies in the Six Day War of 1967 was another pivotal event. The big losers were Syria, whose Golan Heights frontier positions were occupied by Israel, and Jordan, which lost control of the West Bank territories. In 1970 Jordan decisively crushed Palestinian exile groups which were using Jordanian territory as a base of operations; Palestinians—the innocent with the guilty—were driven even farther afield in their exile. Jordan eventually renounced claim to the West Bank in 1988, leaving the remnant Palestinian population in a legal limbo.

The crushing defeat at the hands of Israel threw several Arab states into inner turmoil. Arif's Iraqi government, already suffering from the continuing Kurdish revolt and Communist insurrection, was deposed in a May 1968 coup. Somehow Saddam Hussein reappeared, leading one of the prominent units in the coup. A second coup quickly brought Ahmed Hasan al-Bakr to power, with Saddam Hussein as his deputy.

Bakr had promised continuing primacy in internal politics to the army, but soon reneged on the deal, sending many of the generals into exile. (Saddam, with his typical attention to detail, would eventually arrange to have most murdered.) Saddam became head of the internal security apparatus, the path to power in totalitarian regimes, and like Stalin mastered the art of repression.

In November 1970 the Syrian Defense Minister, Bashar el-Azzad, seized power in Syria, establishing a family dynasty that still clings to power.

Saddam Hussein was growing beyond simple muscle for the regime. As Deputy President he steadily usurped presidential power, driving foes into exile. When the army proved unable to control the Kurds, Saddam bought off their essential foreign support. He negotiated an agreement for massive arms purchases from the Soviets, and in 1975 renegotiated an agreement with the Shah of Iran giving that country effective control over the Shatt el-Arab waterway. Kurdish resistance collapsed within weeks. In the first agreement the Soviets cynically sold out the Iraqi Communist party.

The latter agreement placed the Iraq-Iran border on the Iraqi shoreline, rather than in the *thalweg*, the axis of the river channel. This gave Iran effective control of the waterway, Iraq's only outlet for tanker traffic.

Saddam was also busy on other fronts, finalizing nationalization of the oil industry in 1972; a large proportion of the revenue went to buy those Soviet weapons. Saddam without doubt brought Iraq into the modern age, using oil revenue to promote education, granting women rights unheard of in the Arab world, and in general, Westernizing the society.

The 1973 Yom Kippur War, a coordinated assault by neighboring Arab states upon Israel, was a rarity in that it produced a tangible peace. Jordan declined any direct intervention. The Israelis and Egyptians fought to bloody mutual exhaustion, paving the way for an American-brokered peace, albeit one that was decidedly unpopular in most of the Arab world.

The war was the last of the great confrontations between the Arab states and Israel. It also permanently established a Palestinian refugee population, and set the ground for the disastrous Lebanese civil war of 1975–1990. The war both destroyed one of the most affluent and cosmopolitan Middle Eastern societies, and eventually drew the United States back into the region, with disastrous results.

In 1979 Saddam decided to end the obvious sham of Bakr's presidency. He hankered to become the leader of the Arab world, and feared that an agreement to merge the Ba'athist regimes of Syria and Iraq would thwart his great ambition. In July Bakr went quietly into retirement for "health reasons."

Ghastly purges within the party consolidated Saddam's power. The denouement was worthy of Saddam's idol, Stalin. The secretary-general of the party, who had opposed Saddam, addressed the assembled party officials, describing a bogus Syrian conspiracy. Saddam then named numerous co-conspirators, and invited the assembled party members to participate in the immediate mass execution. The executions were filmed for national distribution. Saddam had derailed the Syrian merger, consolidated power, set a very public example to any who opposed him, and bonded the party hierarchy in a blood-drenched pact.

The party insinuated itself into every aspect of national life. As long as they did nothing to attract the attentions of the state security apparatus or Saddam's family, the masses could live a fairly comfortable life in a welfare state. The cost was absolute and unquestioning obedience. The cost of dissent was immediate and often grisly death.

Still, Saddam wanted to be the leader of the entire Arab world—indeed, the entire Islamic world. Events unfolding in Iran would send him down a far different path.

Domestic resistance to the Shah's rule in Iran coalesced around the Ayatollah Khomeini, who had lived in exile in Iraq and Paris. Khomeini had re-interpreted traditional Shi'a belief that religious rulers should eschew worldly political power, and planned to export his revolutionary brand of Islam to the world. The overthrow of the Shah brought Shia fundamentalists to power, rekindling the old threat on Iraq's eastern frontier. The Iranian militants expended much of their energy on confronting the United States, and the Iranian military, once the most powerful in the region, was in near-total collapse. Still, the massive oil wealth of Iran, and its large population, made it a power to be reckoned with. Khomeini saw through Saddam's attempts at conciliation, and called for his overthrow.

On 22 September 1980, Saddam launched a massive air attack on Iran. The surprise attack failed, and Iranian aircraft struck back at Iraqi airbases, naval bases, and oil production and shipping facilities. Iranian ground defenses proved less resilient. The Iraqi Army drove deep into Iran, but failed to capture key objectives such as the main Iranian oil export facility at Abadan. The Iranian leadership wielded population as a weapon, sacrificing tens of thousands of untrained adolescents in human wave attacks that overwhelmed the better-equipped Iraqis. Iran's ability to interrupt tanker traffic through the Straits of Hormuz struck a telling blow at Iraq's warmaking economy. Disaster was averted by the backing of the United States (still smarting from the seizure and prolonged captivity of its embassy staff in Tehran) who provided limited military support but more importantly in 1984 allowed other nations' tankers to fly the US flag.

Iran was hobbled by shortages of spare parts for its American-equipped forces. Iraqis mastered battlefield engineering, constructing defenses in depth that channeled Iranian human wave attacks into mass killing grounds. Even Iraqi Shi'a conscripts, reluctant to trade secular for religious oppression, fought well. In a final irony, the only active internal resistance to Saddam, in support of Iran, came from the ever-troublesome Kurds. Even so, by 1982 the two countries were locked in a hideous war of attrition.

The Iran-Iraq conflict only reinforced the need for the Marine Corps to plan for multiple types of conflicts, now including mechanized warfare against local dictators flush with armaments purchased with oil, banking, and other monies. The Corps struggled to reconcile the need for adequate fighting vehicles with the ever-present need—and ingrained desire—for "lightness." After several expensive but failed development programs the Marines eventually adopted an "off the shelf" solution, a General Motors–Canada design based on the Swiss Mowag eight-wheeled armored car.

The Light Armored Vehicle (LAV) family seemed to offer everything the Corps wanted: simplicity; it could be airlifted by the CH-53 heavy lift helicopter or carried inside a C-130; multiple vehicles could be carried by landing craft; at least limited amphibious capability; and above all the ability for the basic LAV-25 variant to transport four infantry-scout dismounts—the basic Marine Corps infantry fire team.

The problems with the LAV were those inherent in any light vehicle: thin armor and a small main gun (a TOW-missile armed variant would be provided by mating the basic launcher from the Army's anti-tank variant of the obsolescent M113). The real problem, though, was doctrinal. Despite its lightness mantra, the Corps had no clear doctrine for the use of light armor. Was it to be just infantry support? (At first the units had no dismount scouts, but "borrowed" infantry from supported units). Mounted light infantry with organic fire support? Mechanized reconnaissance? Mechanized scouts tasked with screening and delaying the enemy until heavier, more capable assets could be brought to bear? The troops in the ranks and theorists alike worried over this conundrum, one that was reflected in the evolving designation from LAV Battalion, to Light Armored Infantry Battalion, to Light Armored Reconnaissance Battalion.[16]

Repulsive as most of the world found Saddam's regime, at the time it seemed the lesser of multiple evils. Many were prepared to see him as a bulwark against militant Iranian fundamentalism. In 1983 the Marines were among the first to suffer attack by Iran's new war by proxy.

BEIRUT INTERLUDE, 1983

The second American intervention in Lebanon came about as the result of the collapse of effective government caused by an Israeli invasion. Given the peacekeeping mission, the Marine infantry and tanks were limited to patrolling an assigned zone of occupation, and adding a credible threat to reaction forces.[17] In the end the tanks played little role in the tragedy that ensued. Forbidden by rules of engagement in which any response to attack (even loading weapons) had to be approved from Washington, the Marines were sitting ducks. On 23 October 1983, a truck bomb killed 241 Americans, mostly Marines, and wounded 115 more. For the Marines it was the worst single day's casualties since Iwo Jima. On the same day 58 French paratroopers were killed by another truck bomb.

CARIBBEAN INTERLUDE II—GRENADA, 1983

One of the oldest European colonies in the Caribbean, Grenada's spiral into chaos began well before its independence in 1974. In 1950 Eric Gairy returned from exile, and eventually established a dictatorship. On 12 March 1979 Gairy was deposed in a near-bloodless coup. As brutal and capricious as Gairy had been, things became worse under the systematically brutal Marxist, Maurice Bishop. The Soviets, however, favored the ambitious Bernard Coard, who orchestrated the "Bloody Wednesday" coup of 19 October 1983. Extreme violence caused concern for the safety of foreign students at the St. Georges University School of Medicine, though all factions in the country were anxious to protect the students, who were now the main source of foreign revenue. President Reagan meanwhile seized an opportunity to overthrow a Communist regime. For the US military, the task was formidable: ". . . a large combined force had to be cobbled together at speed," with the added complexity that contingents from several Caribbean nations would be included to legitimize the operation.[18]

As part of an overly complex plan, the 22nd MAU would secure the northern part of the island, expected to be the final redoubt of the defenders. The 22nd MAU was built around 2/8, Hotel Battery 3/10, five M60A1 tanks from 3rd Platoon/A Company/2nd Tank Battalion, and fourteen LVTP-7s.[19]

On D-Day, 25 October, the plan foundered on last-minute changes, weather, the inexperience of the Special Operations units, and its own complexity. By day's end only a few Marines were ashore. In the south, all save one of the Special Ops missions had failed.[20] Hamstrung by communications failures, the Marines still managed to redeploy two rifle companies and mechanized assets to the west coast. The Marines, who had first boarded their assault vessels at 0345hours, landed at Grand Mal, two kilometers north of Government House, at 1901hours.

A task force built around Golf Company started south in the pre-dawn hours of 26 October, with the tanks bringing up the rear. The coastal road was barely wider than the vehicles, and at several chokepoints was perched along hillsides above the sea. Resistance melted away at the squealing sound of tank tracks. That was good, since the tanks had landed with no ammunition for their 105mm main guns.[21] With the tanks providing machine gun support—and intimidation—the Marines relieved the isolated SEAL detachment at Government House at 0730hours. Golf Company pushed on to Fort Frederick, headquarters of the Grenadan defense forces. By nightfall Rangers, transported by Marine helicopters, had collected the St. Georges students at Grand Anse. The following day the only task remaining was to patrol the nearly deserted streets, and to "rescue" a handful of Canadian vacationers—who declined rescue.

Again the Marine tanks had proven effective through sheer intimidation. At the time the chaotic Grenada intervention was hailed as a triumph, and on the whole the Marines had performed better than most. While some units received tumultuous welcomes at their home bases, the Marines boarded ship to relieve the decimated force in Beirut.

————

As the Middle East spiraled ever downward, from Libya Muammar al-Qaddafi exported his brand of international terrorism. On the Iran-Iraq war front money flowed from the royal coffers in the frightened Gulf States to Iraq, and modern arms from the Soviets and others; the United States did not directly provide armaments, only "crop-spraying" military helicopters and intelligence data.

In 1984 Saddam decided to use chemical weapons to repulse a successful Iranian offensive. Following international outrage, Saddam temporarily desisted from their further use. The tanker war escalated when Saddam began to use *Exocet* anti-shipping missiles, in one episode attacking the destroyer USS *Stark* in 1987. The situation was further muddied when it was determined that the impoverished Iranians were obtaining spare parts and weapons from the US in exchange for influence in attempting to gain the release of American hostages in Lebanon, leading to the so-called Iran-Contra Affair.

By early 1988 Iran was on the brink of economic collapse, and had suffered over a million casualties. In August Iran accepted a UN-brokered ceasefire.

Iraq had also suffered severe economic damage. Once one of the strongest regional economies, it was effectively bankrupt. Domestic austerity programs did not sit well with a population that had once enjoyed the benefits of an oil-funded welfare state, and tens of thousands of veterans were unemployed. The Kurds were again seeking autonomy. Under intense diplomatic pressure, Iraq had discontinued the further use of chemical weapons, but in 1987 and continuing into 1988 Saddam directed their use against the Kurds; one attack alone inflicted some 15,000 civilian casualties. Many Western nations were alarmed by Iraq's continuing efforts to acquire nuclear weapons.

CARIBBEAN INTERLUDE III, PANAMA, 1989

The United States continued to be distracted by other regional problems, including the growth of international drug cartels and "narco-terrorists." Problems between Panamanian dictator Manuel Noriega and American forces in the Canal Zone (CZ) continued to fester, and in May 1989 companies from the 2nd LAI Battalion began rotational assignments to the CZ. By 20 December conditions had come to a head after the murder of a Marine by Panamanian security forces.

Fourteen vehicles from D Company (-), 2nd LAI, with unfamiliar dismounts drawn from a security company, took part in Operation JUST CAUSE, the overthrow of Noriega. Task Force SEMPER FI quickly secured the western approaches to Panama City and the Bridge of the Americas spanning the canal. Later the task force's LAVs were used to reduce roadblocks and provide fire support.[22]

Iraq was determined to recover the cost of its war with Iran, which Saddam saw as one fought for the benefit of all Arab states against the Ayatollah's Shi'a regime. In a period of falling global oil prices, members of the Organization of Petroleum Exporting Countries (OPEC) began to exceed their sales quotas, further damaging Iraq's economy, whose only other export was dried dates.

Instead of forgiving Iraq's war debt, Arab governments began to press for immediate repayment. The leaders included the United Arab Emirates, but the most foolhardy was the Emirate of Kuwait. Despite sharing a border with a nation that had the largest and most battle-experienced army and air force in the region, the tiny nation demanded repayment. In July 1990 Iraqi diplomats grew increasingly bellicose, accusing Kuwait of stealing oil by increased production from the Rumalia oil field which spanned the national border.[23] Even King Faisal of Saudi Arabia urged the Emir to give in to Saddam's demands. Finally on 25 July Iraq delivered a letter to be delivered to President George H.W. Bush that was essentially an ultimatum. The American ambassador to Iraq, April Glaspie, seemed blind to Saddam's message.

On 2 August 1990, three Republican Guard divisions spearheaded an assault that swept aside the tiny 16,000-man Kuwaiti army. The royal family, along with elements of its military, civilians, and foreign workers fled the country.

The invasion sent an economic shock through the industrialized world. The powerful Iraqi Army threatened half the world's oil reserves. Saddam had sorely miscalculated global reaction, and could no longer depend upon his last, now-tottering patron, the Soviet Union. After a series of increasingly stringent resolutions, embargoes, and a naval blockade failed to dislodge the Iraqis, the UN on 29 November authorized Resolution 678: the removal of Iraqi forces from Kuwait "by all means necessary."

NOTES

1 For example, see Keegan, *The Iraq War*, p. 8–55.

2 Ibid, p. 15.

3 Ibid, p. 15. The Marsh Arabs are a minority population occupying the wetlands at the mouth of the Tigris-Euphrates delta. Long intractable, they are now best remembered as the victims of Saddam Hussein's "final solution," draining the tidal marshes to eliminate their traditional homes, refuges, and way of life.

4 Faisal, the son of the Sharif of Mecca, was a Sunni raised in Istanbul.

5 The Druze faith is an offshoot of Islam which incorporates aspects of several other philosophies, and is sometimes considered a separate religion. Persecuted for centuries, and isolated in mountain domains, like the Ghurkas they have often proven loyal mercenary warriors.

6 For a comprehensive history of Middle Eastern oil exploration, discovery, and energy politics, see Daniel Yergin, *The Prize: The Epic Quest for Oil, Money and Power.*

7 One casualty of the fighting was Iraq's sizeable Jewish community. An integral part of the culture since Assyrian time, many were killed in the first of the nationalist *farhud.*

8 Shihab is transliterated phonetically from Arabic. In some sources it is spelled Shehab or Chihab.

9 As quoted in Alasdair Soussi, *50 Years Later, U.S. Marines Remember The 1958 U.S. "Intervention" in Lebanon.*

10 Craig Symonds and William Clipson, *The Naval Institute Historical Atlas of the U S Navy,* p. 200–201.

11 Estes, *Marines Under Armor,* p. 159.

12 Ibid, p. 159. Marine tank platoons could be operationally subdivided into a heavy section of three tanks under the platoon officer, and a light section under the senior NCO.

13 For a more detailed summary of the lengthy involvement of the Marine Corps in Dominican affairs, see Millett, pp. 137, 179–183, 194–207, and 556–558; and Ringler and Shaw, *U.S. Marine Corps Operations in the Dominican Republic, April–June 1965.*

14 The suffix (-) indicates less than the total unit, (REIN) indicates reinforced. In a tank company (REIN) typically indicates support such as maintenance or communications elements from the battalion H&S Company.

15 For the Marine Corps armor involvement in that most frustrating of wars, see the companion volume *Marine Corps Tank Battles in Vietnam.*

16 Estes, *Marines Under Armor,* p. 180–182; Michaels, *Tip of the Spear,* p. 3-4.

17 Turner, *Tanks with the MEU: A Team for Success,* p. 39.

18 Ibid, p. 39.

19 Ibid, p. 234–235; Estes, *Marines Under Armor,* p. 182.

20 Adkins, *Urgent Fury—The Battle For Grenada,* p. 251.

21 Estes, *Marines Under Armor,* p. 182.

22 The company included LAV-25, LAV-L, and LAV-C2 variants. Panamanian armor included V-150 and V-300 armored cars, the latter armed with a 90mm cannon. Varying numbers of LAVs available are cited by different sources. Estes, *Marines Under Armor,* p. 183, Reynolds, *Just Cause: Marine Operations in Panama, 1988-1990,* p. 15, 22–26; Johnson et al, *In the Middle of the Fight,* p. 131–133, 135.

23 Arabic is often translated into English in several different forms. Alternative spellings include Rumeila, Rumailia, Rumalla, and others.

CHAPTER TWO

Operation Desert Shield

"I don't know what effect they will have upon the enemy, but by God, sir, they frighten me."—Arthur Wellesley, Duke of Wellington, observing his own troops

T HE IRAQI SEIZURE of Kuwait provoked an immediate response from the Western world, and caught the Marine Corps with units deployed from Sierra Leone to the Mexican border.[1] The first American unit on scene was the 7th Marine Expeditionary Brigade (MEB), a desert-trained unit with three infantry battalions and an LAI battalion.

The first tank unit to deploy was LtCol Alphonso Buster Diggs' 3rd Tank. Diggs was at Chico State College when an officer selection team was confronted by anti-war demonstrations. He fell for an old ploy. "When it was all over the protesters all left. I just picked up the literature, and this Captain, he said 'What are you doing there son?'

"I said 'Ah, I'm just looking at the literature'

"He goes 'Don't worry. You couldn't qualify for this program anyway. It's too hard.'"

Diggs ended up in Platoon Leaders class between his junior and senior years. His initial assignment was in amtracs, and he attended the Advanced Armor School at Fort Knox. "There was a guy there named Paul Lessard. If he liked you, he would put in for an MOS change to tanks, because both amtrackers and tanks went there [as] captains."

Diggs—like most of his officers—was new to the battalion. Rick Mancini of Alpha Company had been a college student when he casually talked to a Marine officer about the Marines and ended up in the officer candidate program. When duty assignments were announced "I probably made half the class pissed off at me because of the fact that so many people wanted tanks and didn't get it, and I didn't want it and got it." At Fort Knox Mancini found his calling thanks to an outstanding Army NCO, Staff Sergeant Layton. Assigned to 3rd Tank, Mancini had been aboard only three months.

Diggs's tank commander was Sgt Kevin Kessinger. Kessinger joined the Reserves to help pay for college. Not liking college, he volunteered for active duty and "ended up loving it."

The battalion commander could direct his unit from his tank (Hotel One), an AAVC-7A1 command track, or Humvee as he saw fit. Diggs preferred a tank, but the problem was task overload. He had to direct his own crew, monitor several radios, direct the battalion, and coordinate supporting arms. Diggs solved the problem by riding as loader, letting Kessinger control communications and the tank. "Essentially we were a three man crew." At stops "Colonel Diggs hopped off the tank and didn't do the kind of duties a loader would do. Really your gopher guy, does a lot of the menial tasks on a tank. Colonel Diggs wasn't about to do those things."

The tank was modified for Diggs. Kessinger: "My job was always to get the task force tacs [radio frequencies], the regimental tac, battalion tac, even some of the company tacs, have them pre-loaded on the RT246 so that the colonel, all he had to say was who he wanted to talk to. I kind of kept the channels organized so depending on who he wanted to talk to, I could put him on the right channel. . . . We kept his maps up to date, and made sure he had as much information as we could put in front of him. . . .

The other members of the crew were the driver, LCpl Kevin Moroni, and gunner Galvan. Kessinger was particularly proud of Moroni, who had come as a "messed up kid," but became an excellent Marine.

Mike Mummey had grown up with his uncle's stories of tank warfare in World War II. In 1977 a Marine Corps recruiter had been calling a friend, and Mummey went with him to "tell the recruiter to leave him alone. As soon as I walked in there's a picture of an M-48 going through a big mud bog. I looked up at the staff sergeant and said 'You guys have tanks?' He's like 'Well yes we do!' That was it. Next thing you know I get my parents down there to sign for me because I was only seventeen."

But "When I got out of boot camp I had orders to go to Twentynine Palms to be (MOS) twenty-eight-hundred communications Marine. When

I talked to the series gunnery sergeant there, the Chief Drill Instructor, he said 'Hey, shithead, that's because they have radios in them tanks. That's why you're gonna be in tanks.' That wasn't true.

"When they found out I wasn't supposed to be a twenty-eight-hundred they said 'Well, you can go to ITS (Infantry Training School), everybody can be a grunt.' So I went to the grunts."

Mummey was discharged in 1979, but re-enlisted in 1984. After a tour as a Drill Instructor he joined 3rd Tank in 1988, and became the Platoon Sergeant of 1st Platoon, Charlie Company. Third Tank was training for desert warfare. "We were always out in the field. My first year at Twenty-nine Palms we were out in the field 283 days."

In modern war the logistical effort is paramount, and the combat support units—engineers, logistics personnel, cargo handlers, and others—are often the first to arrive. Normal procedure would have been to deploy these "off load personnel" to Diego Garcia to travel aboard the ships, but the ships had already departed Diego Garcia. Combat personnel were flown to Kuwait aboard chartered airliners to marry up with equipment loaded aboard the ships of Maritime Pre-Positioning Squadron 2 (MPPS-2) from Diego Garcia. But first logistics specialists had to unload mountains of weapons and materiel.

Chris Freitus, Executive Officer of Charlie Company, 3rd Tank described a long flight packed inside the airliner, not allowed to disembark for thirty-six hours.[2] The temperature was 106F (41C) when the plane arrived at 0200hours. Buster Diggs thought conditions were not bad "Except humidity. Twentynine Palms is dry. Remember that's on the [Persian] Gulf. It's horrible. When I opened the door of that plane, I said 'I'm not gonna be able to do this. I'm forty-two and I'm not gonna be able to do this.' But you acclimate quickly."

Mummey:"We get off the airplane and operation chaos begins because it's going so quickly there's not a lot of liaison people that went over before us. There's a lot of big question marks hovering around. You definitely had to go and find out what was going on." The Marines boarded trucks and chartered buses for the ride to the port at al-Jubayl. The trip was an introduction to local driving practices: mutter a prayer of "Insh'Allah" ("God willing"), and do something no sane person would attempt.

The Saudis confined the Marines to the immediate areas of the piers, over 9,000 Marines crowded into warehouses with inadequate water supplies and few functioning toilets, in temperatures above 120degrees F (49C). "Those next eight days were the most miserable days of my life, easily," recalled Kessinger. "We got put into these huge warehouses that lined both sides of the pier at Jubayl as we linked up with our MPPS. As we were waiting for the ships to come in, we were asked to stay inside because we didn't have air superiority yet."

Several thousand men were "pretty much just lying around the inside of this football field-sized warehouse, just literally lying on cement floors. It was a hundred fifteen, hundred and twenty outside, so it was hotter inside. I remember lying on a piece of cardboard, panting like a dog. Uncontrollably." The men spent eight days in the warehouse.

Buster Diggs: "None of those things work as smoothly as [they're] supposed to work. FSSG is supposed to fly in first, you fly in five days later, FSSG has the equipment all ready for you. That didn't happen because we needed the troops on the ground quickly."

Kessinger: "When the ships came in and there was work to be done, that was better. We did that in shifts. Those [FSSG] guys took care of us the best they could with what they had. I remember just cool water was a luxury."

"We drove just about anything off the ships," sorting through the gear on land.

Rumors were rife. Mummey recalled that "There were supposed to be Iraqi special forces frogmen under the pier, this and that. There was a lot of bum scoop going around."

The tankers found that equipment necessary to operate the tanks—tools, fluids, maintenance equipment—was missing. The crews bartered; an icechest for a set of wrenches, tool boxes for batteries. Douglas Smith was a Navy Reserve corpsman: "We've had to fight for everything. We almost stole the tanks off the ships in order to get them."[3]

Part of the problem was sheer bad luck. Buster Diggs: "Let me put this in perspective now. Our [MPS] Squadron was due for maintenance. In fact, one of our ships was in Blount Island [Florida] for maintenance. The contractor crews had flown off. The stuff was probably in a poor state of readiness only because it was going back to Blount Island for maintenance."[4] "They had the wrong weight oil. You don't know where you're going to fight. You gonna fight in the desert? So that's kind of unfair."

Tanks were missing gun sights, batteries, rangefinders, and other parts. "There was a yellow IOU tag hanging there," said Mummey.

A major problem was the reactive armor on the old M60s. The Marine Corps had aborted the upgrade in 1989, and the Navy objected to the explosive tiles stored on its ships. The first shipment of tanks from the US went without the explosive blocks. The explosives were dug out of storage, more were acquired from Army stocks, and all had to be hurriedly shipped and installed.[5]

"The tanks that came off the MPS ships came in a whole bunch of flavors," Mummey added. "One squadron they were all sand-colored, and they all had reactive armor. There's another squadron came in, and all these tanks were in four-color MERDC camo 'red desert,' and had no reactive armor. The tanks that 1st Marine Division—1st Tank Battalion—drew were three color black, green and brown NATO colored, with reactive armor. Later on these tanks from Okinawa started showing up, and also from Lejeune . . . they were four-color jungle MERDC, some with reactive armor and some without." Teams from the Anniston Army Depot were flown in to Jubayl. "They would put on the antlers (brackets) as we called it for the reactive armor." The crews had to add the actual explosive tiles, "And that was a pain in the butt, because normally you take a dummy tile off and put a live one on. These tanks would come up, and you wouldn't know which way [to put the tiles on]. You had to park it next to another tank and use it as an example."

With more experience at deployments, the LAV personnel brought aboard their civilian transport planes much of the equipment missing from the stored LAVs: maintenance tools, machine guns, and secure SINGCARS radios.[6]

The Marines discovered that at night dew gathers on every surface, and combines with salt dust in the air to corrode exposed metal almost instantly.

The Marines were frustrated by the lack of cooperation from local officials. It seemed that some prince owned the local concession for nearly everything. In perhaps the sandiest country on the planet, sand to fill sandbags had to be purchased.[7]

By 25 August, ten days after arrival, 7th MEB was ready to defend the port. Even as the tanks deployed forward to become a "tripwire," the very basics were in short supply. Mummey: "When we rolled out all I had was one can of two-hundred rounds for my coax machine gun, and whatever nine millimeter of five-five-six we took from a pallet at the Air Force Base in Riverside before we flew. . . . The ammo in the MPS hadn't been discovered yet." The Marines quickly concluded, "That was George Bush's big bluff. He put us out there in the sand. We didn't have ammo: no main gun, we had like two missiles. That was it."

Kessinger explained: "Third Tanks was in Camp Five. We rotated at about fifty per cent strength from the field back to this Camp Five. I think we did six days out, three days back or something like that. We were constantly moving the battalion around in the different positions." Inevitably some platoons were back at the base camp when the battalion moved. Mummey: "You'd have to move the other platoon's tanks. That turned out to be a real adventure." The men had to pack up gear and shift multiple tanks with short crews. "That included taking down that big circus tent net, folding it up. The battalion would do its road march, we'd move to a new battalion assembly area two or three clicks in diameter. You'd have to set up the tank, then set up their net. Go back over to your tank" and repeat the process. The tanks moved at night to avoid inconveniencing the local drivers.

James Gonsalves, the son of a Navy corpsman, was impressed by the Marine faculty members at the Naval Academy. He graduated Tank School in September and in late October joined 3rd Tank already in Saudi Arabia. Charlie Company commander Ed Dunlap drove him ". . . right to my tank platoon. Said 'Okay, talk to the platoon commander, and do a quick turnover. The platoon's yours. Don't mess it up.'" The outgoing leader "Basically said a quick hello, jumped into the Humvee with the company commander, and he went off to his next job, which was a TOW platoon commander. . . . I turned around and there were fifteen Marines staring at me, and it was a pretty humbling experience."

At this point the Marine armor force consisted of 53 M60A1 tanks, 28 LAVs, and about 100 AAVs.[8]

Freitus discovered that confusion was even worse for some units. About mid-day a bedraggled convoy of Army airborne armor pulled in from the airbase, looking ". . . totally lost, shell shocked." M551 light tanks were running short-tracked, and one was towing a disabled M113 personnel carrier. The Army platoon commander needed parts and supplies, even water. Unable to offer any mechanical assistance, the Marines gave them food and water, and pointed them toward where they thought Army troops might be.[9]

More units of I MEF arrived within days, the leading edge of a force that would grow to two reinforced divisions, an enlarged air wing, and two service support groups.[10] Two MEBs remained afloat as an amphibious threat the Iraqis could ill afford to ignore.

Shoulder holsters, supposed to be an issue item for tankers, were purchased with personal funds. Other items as mundane as sunglasses were donated, though in many cases such items found their way into the hands of others along the long supply chain, and never reached the front-line troops.[11]

The Marines and airborne units would have to defend the border for weeks before heavy Army units could arrive. They referred to themselves as speed bumps. Diggs: "We weren't ready. We knew what was up there. Jeez, I think they had three tank and five mech divisions in Kuwait. I was the only tank battalion there. I wasn't ready to attack. That's ludicrous." Saudi units formed a thin warning line, with the Americans behind. In theory they would trade space for time, with air power and tank-hunter teams trying to slow any attack. But there was no space to trade. In case of invasion the Marines would have to fight in place. Major General John I. Hopkins, the commander of the 7th MEB:

> We would screen as far forward as possible, delay and attack the Iraqis with air power, then defend in a main battle area along what became known as "cement ridge." The Iraqis had two possible attack routes. We thought they'd either come down the coast or use a route a little bit to the west, but both these routes come together at a junction near the cement factory. If they kept coming, we had drawn a line in the sand by the cement factory. We were going to stay there.[12]

Of their defensive position on Cement Ridge, Diggs recalled that "It's high ground, probably twenty meters higher than every place else, and they dig in. Task force commander says 'I want overhead cover' and all this crap. I said 'Jeez, they're just gonna drive around us.' Why would they attack down this road, right into our teeth, when they could drive out in the desert and go around us? So I stayed mobile. I thought it'd be a meeting engagement out in the desert."

Like many, Hopkins was also concerned about facing the more modern T-72 tanks with old M-60s.

The only thing we've got to do is when they come, we've got to close with them right away and take away the advantage they have of outgunning us. In close, we'll have more maneuverability, we'll have the sabot round, and we'll cause some problems.[13]

The mixture of old and new vehicles, and the enormous requirements, caused logistical problems. Dennis Beal was Buster Diggs's S-4. "Fuel was a big issue because you brought your five-thousand-gallon refuelers with you . . . you need two hundred thousand gallons of [in] bladders on the ground to keep this so you can fill up your refuelers, you can fill up your tanks, you can fill up your wheeled vehicles that are flowing back and forth. Then this bladder farm, which many of them were over a million gallons, that has to be constantly refueled. That to me was the biggest issue. MREs coming out the yin-yang, there was no problem with food. Water wasn't an issue. You could fly in pallets of bottled water, but you don't just make fuel on the spot."

The M60A1s experienced fuel problems because CENTCOM had adopted a single-fuel policy, and the available high-quality fuel caused degraded performance, high fuel consumption, and clogging of the fuel system. The tanks had been operating on diesel fuel, and had condensate water and impurities. When jet fuel was added "That immediately started going to the fuel cells, and we ran out of fuel filters" said Mummey. "They'd barely go three miles an hour. I saw some creative stuff. We'd pull the fuel filters out and they'd blow 'em out with an air gun, wrap them in a sock . . . and stick 'em back in there and get it down the road." Parts arrived by odd routes; in one instance Mummey's crew received a FedEx box addressed to the tank. "We all thought we was gonna get a special present. We pull it open and it's just a bunch of stinkin' fuel filters."

Fuel also caused problems with the LAVs, designed to operate on commercial grade diesel. The fuel purged accumulated sediment and contaminants from the fuel systems, clogging fuel filters and injectors.[14]

Tank battalions using the M60A1 also received a tiny handful of more modern M60A3 tanks. Buster Diggs: "The first Army tanks that were over there were M60A3s, and as soon as M1s came, they abandoned those and we picked up some of those, too."

The problem in using two marks of the M60 was the thermal sights. "The tank automotively is the same tank. The 'A3 had a thermal sight device, night passive vision. We liked it. The 'A3 had a better sight than the M1A1. That thermal sight was tremendous." Mancini acquired two of them, and was saved by his First Sergeant who had been to Master Gunner's School, and "That was a real blessing."

Marine deployment of the M1A1 tank had been delayed for budgetary reasons. Tankers were hastily trained on the new M1 series tanks before flying to Saudi Arabia, where 16 M1A1 and 60 new M1A1(HA) tanks were issued from Army stocks to the 2nd Tank Battalion.

Two companies of the Reserve 4th Tank Battalion acquired the new tanks. Bravo Company of 4th Tank was not activated until 17 November. They departed for Twentynine Palms on 15 December for two weeks training on the new tanks; the training course is normally eleven weeks.

Although the Reservists of 4th Tank had worked long and hard to achieve high proficiency, they met with the scorn often accorded "weekend warriors." The major in charge of the school informed company commander Captain Ralph F. "Chip" Parkison that he was not pleased at dealing with Reservists. The major assured Parkison that his men would not satisfactorily complete the intensive training. Parkison replied that "With all due respect, sir, you are in for a big surprise."[15]

Formal training lasted until 12 January, conducting both day and night firing, and exercises over the rugged Mojave Desert. The Reservists completed the training with the highest proficiency ratings.[16] As a Reserve unit equipped with the Corps' newest and most expensive weapons system, there would be many eyes fixed upon Bravo Company.

With few of the new tanks in the supply line, Alpha Company, 4th Tank would go to war in M60A1s. A recruiter visited Samuel Crabtree's high school, and "Everybody said the Marines were the hardest, so I said 'Well, I might as well try that.'" His father—former Air Force—advised him to join the Reserve: "If you like it, then you can go active duty."

Crabtree thought that the M60s ". . . were easy to maintain. They weren't very complicated. . . . Once you got them running, they pretty much ran non-stop." The biggest problem was "If the starter goes out, you have to remove the whole power-pack, just to get to the starter. Something as simple as a starter could take many hours, and you have to use the Eighty-Eight in the field. . . ." They developed a workaround for the fuel problem. "The sixty wasn't designed to handle multi-fuels. What you had to do was add oil to the fuel. . . . The problem was it would clog up the filters if you put in too much."

They took on M60s from those left behind by 1st Tank, and sailed on the USS *Tarawa*. The Navy allowed the tankers to run the engines infrequently to charge batteries. At a stop in the Philippines the tanks were disembarked ". . . to shoot them and drive them around. The problem we had there was the fuel cells were made out of metal. The age of the metal, they started cracking. We had to pull them all out, rinse them out, weld them, and then put them back in."

LAV units were now "Light Armored Infantry," with organic dismount scouts. The composite unit was an amalgam, with Alpha and Charlie Companies from 1st LAI, Bravo and Delta Companies from 3rd LAI. Internal coordination was a problem exacerbated by differing training, procedures, and even doctrine from one battalion to another. The battalion commander, Col. C. O. Myers, developed a whole new unit identity, beginning with designation of the unit as Task Force SHEPHERD.[17]

Particularly troublesome were the inability of the LAV-25s main gun to penetrate tank armor, and the absence of a thermal sight. The LAV-AT's outdated missile launcher system took about 30 seconds to deploy, and up to four minutes to re-boresight the weapon after each time the vehicle was moved.[18] The Marine Corps also has to rely upon the AAV as its primary infantry carrier. The fleet of AAVs aboard the MPPS fleet lacked key updates, particularly to the transmissions and weapons systems.[19] In Saudi Arabia there were no training areas for heavy weapons or integrated live fire exercises. For tank crews, the only truly intensive training was in obstacle breaching.[20]

Christopher Swift, a new crewman in Charlie Company/3rd Tank, said most training centered on forming the battalion into a defensive coil, and putting up camouflage nets. The dark green camouflage nets were of dubious value in the desert, but Buster Diggs felt that any small advantage the nets could give might be critical. "I was not sure we weren't going to be attacked. . . . If we're going to be attacked, I feel sorry for my brother battalion commanders, but it aint gonna be Third Tanks!"

Greg Michaels found that the nets were useful; erected on a barren hillside or in a ravine, they could be used to lure "opposing forces" into an ambush. More effective was "desert paint," local mud smeared over the vehicles.[21]

Eventually the Marines were allowed to construct improvised firing ranges, only to discover that the desert was not that empty. Bedouins had to be shooed out of impact areas. Task Force SHEPHERD constructed targets, but scarce building materials were scavenged by other units. The battalion acquired wrecked vehicles to serve as targets. Sergeant Ramirez of the battalion headquarters was determined to guard these prizes and slept in the desert to protect them. One night Ramirez listened to reports of an Iraqi attack, and spent a nervous night hiding near the wrecks. The attack turned out to be a large group of deserters.[22] The tanks were eventually allowed to fire practice rounds and boresight their cannons on measured ranges.[23]

Heavy emphasis was placed on navigation, but there were few landmarks in the low, rolling terrain, and road patterns were confusing. Maps were in short supply, and of dubious value. This was alleviated only when a British cartography unit showed up "And then's when we started getting good maps that we didn't have to make ourselves," said Mummey.

Greg Michaels observed that the desert ". . . was like a living being; it changed every day."[24] Dunes moved with the constant wind, and sand might encroach over roads. Near the coast were *sabkha*, flat plains with thin crusts that overlay stinking brine-saturated mud. Men on foot could break through the crust, and they were completely impassable to vehicles. Thickness of the crust fluctuated with seasonal tides, and a vehicle would break through the crust of a *sabkha* that had been passable days before.

"After a while you could kind of see them," said Diggs, "we were there six months. As soon as you bog down, you don't stop. Don't keep driving forward, fool. Don't do that! It's not gonna get better. Stop and back up. You don't know where the *sabkha*'s going." Windblown sand might form small dunes above the *sabkha*, making it indistinguishable from the surrounding desert.

The *sabkhas* provided the LAV crews with practice recovering mired vehicles using the new kinetic-energy rope. The nylon hawser could be hooked to a mired LAV, and coiled on the ground. A second vehicle hooked up and backed away at speed. The elastic hawser snatched the stuck vehicle free and saved hours of digging.[25]

Considerable ingenuity was devoted to creature comforts. When tanks were retrofitted with turbochargers Kessinger said, "We would take these [old] blower motors . . . and we would hook up a fifty-cal barrel cleaning rod, and we would put a one liter bottle of water into an issue sock, cover the sock with water, and . . . spin two of them at one time, like a centrifuge, cooling the water off. Eighty degree water tastes like ice-water when it's a hundred and twenty."

In the daytime the troops sat in the baking sun, tormented by hordes of flies that swarmed onto food, eyes, human waste, and sweat. More serious threats were dung beetles that crawled into Lister bags seeking water and causing dysentery, scorpions, and spitting cobras. Even the roving camels caused problems as they ate anything in sight, and learned to spring open the top hatch on a water trailer and stick their heads inside.[26]

More pervasive pests were those fans of war, rodents. Molly Moore, a reporter for *The Washington Post* wrote:

MREs developed a major following among the desert's rats and mice. During a stay at one Marine supply center near the Kuwaiti border before the ground war, half a dozen large rats invaded our tent nightly, waking us as they gnawed through the brown plastic pouches and nibbled their way through the contents.[27]

Task Force SHEPHERD established headquarters at the Chicken Ranch, an abandoned poultry farm with a well that supplied stinking, sulfurous water. The troops grew accustomed to the smell of tons of manure baking in the heat. The farm was re-occupied with the coming of cool weather, and the owners brought thousands of baby chicks that peeped day and night.[28]

Kessinger recalled that "In the morning I'd get up before everybody, start the coffee. . . . I'd go to the battalion briefing which was at seven-thirty every morning. Then I would come back, and by the time I got back from that meeting my men were up, and dressed, and they had had chow. We usually had hot breakfast. . . . Once we got to where things were just routine, you had hot breakfast and hot dinner, and we had MREs for lunch."

But "Lots of times we wouldn't eat lunch. Just too damn hot. . . . We would relax at about thirteen hundred, and then we would have some kind of PT in the hottest part of the day, but we would play Nerf™ football, or go for a little run, or do something fun. Then at fifteen hundred we would always take a makeshift shower."

Each crew was alloted twenty gallons of water per day. "We would save ten gallons. We made a makeshift shower on the back of our tank. We took an ammo crate lid for main gun rounds, and on top of that we affixed a fifty cal ammo can where we punched some holes in the back of it. One guy would be standing on the tarp on the ground below the back deck of the tank, and we would fill the ammo can up with water, and rinse off and lather. Then they would fill it up again and rinse off the soap. Then we would take turns. Taking a shower at fifteen hundred after the heat of the day really made things a little bit more bearable.

"Then chow showed up around sixteen hundred, sixteen-thirty. Then after evening chow we would stand around and listen to the Armed Forces Radio. It was always a high point of the afternoon, listening to what was going on. Then we'd spend the night, guys playing chess, writing letters, playing spades. . . . The usual stuff."

Mike Mummey did not let his platoon have cots because "If they had a cot, they tried to build an apartment out of it."

British armored formations trained with the Marines, and the exchange of rations was a welcome change. Less pleasant was a shipment of green apples. The tankers of 3rd Battalion fell upon these unexpected treats, which produced dysentery and dehydration. Chris Freitus: "Once stricken, we soon ran out of toilet paper and developed a routine: roll out of your cot, run into the desert, evacuate from both ends, clean up, bury the stuff, and return to your cot and lie back down completely drained." Much of the battalion was stricken, with some requiring intravenous fluids.[29]

Diggs said his battalion was spared worse afflictions because "I had a battalion surgeon by the name of David Whitehurst. . . he read up on what would happen in a situation like that. He also read that they use night soil [human excrement for fertilizer]. He ordered—I don't know the drug—a big blue pill. . . . There's never any salad. One day they brought out all kinds of salad, and I just chowed down on the stuff. It was local stuff, and they fertilize with night soil.

"Your whole battalion can really be down with diarrhea, the entire battalion could be just on its butt. If he hadn't ordered those pills, we'd have been down on our butts." Whitehurst had ordered so many doses that the battalion supplied other units.

The Iraqis constructed defenses that had worked well against the Iranians, deep "boxes" of open terrain divided by high earthen berms. An attacker was forced to breach or cross each berm, only to be exposed to crossfire on the far side.

Retired LtGen Bernard Trainor briefed them, and Diggs said that "It was the most important briefing I ever got." Trainor told them "They're setting up the same way [as] for the Iranians.' When he explained their system, I said, 'The big difference is the Iranians were walking. I'm not gonna walk. Speed would be our greatest asset. I turned to the infantry [commander], my boss, and said, 'Well, if the infantry ever get out of the 'tracs, we're not gaining ground here.'"

Planners abruptly changed the Marine Corps mission: though amphibious force would remain as a threat, it would be a land campaign. The amphibious ruse tied down four infantry and two mechanized Iraqi divisions, and about half of all Iraqi artillery.[30]

Winter cold and rain brought its own miseries. Cannon ammunition is sensitive to moisture, and machine gun ammunition corroded so that belts had to be cleaned—round by round, and link by link.[31] Mummey said that after Thanksgiving "They sent our supply guy—he flew back to Twentynine Palms and went into our battalion supply warehouse and embarked with some other Marines all of our sleeping bags."

Coalition aircraft began an aerial pummeling on 17 January 1991. Iraq retaliated by lobbing ballistic missiles into Saudi Arabia and Israel, forcing the US to divert air assets to the "Great SCUD Hunt." Diggs underestimated the psychological effect of chemical weapons until one night he observed another officer. "One of the battalion commanders . . . he broke out his MOPP suits. There was obviously a real fear of gas. I just didn't have it. Gas is not as bad as you think if you're moving. If you're stationary it can be a sinofabitch. Just keep going. Get out of it."

The aircraft carrying Reservists from B/4th Tank landed at al-Jubayl just as the all clear sounded after a SCUD alert. Assigned to support the 2nd Marine Division, the company found that new tanks from Army stores were missing critical gear. Five days were consumed in bringing the tanks into operation.

The LAV-ATs and LAV-Ms of the Weapons Company could not effectively communicate with the line companies. Problems surfaced during a mission on 21 January when Alpha Company, Task Force SHEPHERD was ordered to extract a threatened observation detachment from OP6.[32]

The Iraqis had driven the recon team from a cluster of buildings into a rally point in a culvert, abandoning a Humvee and sensitive equipment. The anti-tank and mortar LAVs became separated. When the LAV-25s of A Company arrived many of the recon team leapt into the rear and beat a hasty retreat.

The LAVs and recon men were sent back to retrieve the abandoned equipment. The plan was for 1st Platoon to move into the cluster of buildings while 2nd Platoon provided a base of fire. The mortars were not in the plan, but commenced firing on the objective. Unable to directly coordinate with the mortars, 1st Platoon moved uneasily into the impact zone, risking friendly fire, point-blank RPGs from any Iraqis, and incineration by gasoline from a wrecked filling station.

The recon team recovered their gear and retreated. In the confusion LAV-ATs had moved past the buildings onto the Iraqi side, and did not get the order to withdraw. Greg Michaels had to take his vehicle back amid the buildings and order the TOW vehicles to withdraw.

Commencing on 20 January, the Marines conducted artillery raids. One or two batteries of howitzers, protected by a screen of LAVs, would displace to a point in the desert south of the border. The artillery would fire a hasty barrage against pre-selected targets, and move quickly to the rear. This would inevitably draw counter-battery fire, revealing the positions of the Iraqi artillery for aerial attack or counter-battery fire.[33] Altogether the Marines would conduct a dozen such raids.

Bravo Company of Task Force SHEPHERD conducted another raid on 25 January to destroy a police station used as an Iraqi observation post.[34] The FAC accompanying the LAVs moved up onto the berm, using his MULE to mark targets for air attack, and the artillery opened fire. When the Iraqis returned machine gun fire the FAC retreated. The supporting

LAVs tried to return fire, but their rounds impacted on the friendly side of the berm.

The raid was a success, but at about 0525 a fatigued LAV driver dozed off, and awoke to see a vehicle in the parallel column directly beside him. Speeding up to recover his proper position in the staggered columns, he collided with the rear of another LAV, killing three Marines and injuring four.

TRAGEDY OF THE LIGHT ARMOR: RA'S AL-KHAFJI

Tiny Ra's al-Khafji, a cluster of abandoned buildings around a pier, a desalination plant, and a refinery, was no place anyone would have chosen to fight over. North of a broad expanse of coastal *sabkha*, vehicle traffic was channeled along a narrow coastal highway. Considered indefensible, it was screened by numbered observation posts that overlooked a border berm.[35]

The observation posts were intermittently occupied by reconnaissance and observation units, and screened by companies from Task Force SHEPHERD and the 2nd LAI Battalion. Saudi, Qatari, and Moroccan mechanized battalions lay south of the outposts.

On the night of 28 January observation teams reported enemy mechanized movements. Coalition commanders paid little attention to LtCol Richard M. Barry of 1st Surveillance, Reconnaissance, and Intelligence Group who warned that "the Iraqis want Khafji."[36]

With limited transport, the Marines planned for a "slingshot logistics" offensive. Huge dumps were established ten kilometers forward of the main defenses so that combat elements could move forward unimpeded, and resupply convoys would have shorter trips as the battle unfolded. But the logistical units and massive piles of supplies and ammunition were thought vulnerable.[37]

On 29 January the outpost line was held by Recon platoons and a screen of LAVs. At about 2115 a brigade of the Iraqi *5th Mechanized Division* advanced on positions held by Alpha Company (REIN), 2nd LAI Battalion near OP1, and aircraft disabled a T-62 tank. Shortly before midnight the Iraqis broke off the fight.

The heaviest fighting occurred near OP 4 north of The Heel, the point where the national boundary turned north after extending in from the coast. OP 4 was occupied by a platoon from the 1st Reconnaissance Battalion, supported by Delta Company (REIN) of Task Force Shepherd, with 19 LAV-25s and seven LAV-ATs northwest of the outpost. Charlie Company held an extended line behind OP 6 to the north. Alpha Company was in a reserve position 10km behind the two outposts, while Bravo was 25km to the southwest.[38]

Around 2000hours the Marines spotted a large Iraqi force arrayed in a skirmish line of tanks followed by BMPs. Captain Roger L. Pollard of D Company shifted south to support the Recon detachment and gain maneuver space. Battalion was unable to hear Pollard's transmissions because of Iraqi jamming.

At 2100 hours the *6th Armored Brigade, 3rd Armored Division* struck the Recon platoon. Unaware that the Recon Marines had their own transport vehicles, at 2144hours Lt. David Kendall's 2nd Platoon from D Company, with LAV-ATs accompanying, rushed up to cover their withdrawal.[39]

Kendall:"We were moving forward so that we could get the ATs in position so they could observe better what was going on down there, and understood that once we got in position where they could see the enemy good enough to take a shot at it we'd stop." Unable to fire on the move, the LAV-ATs advanced in bounds. "We stopped on line and the ATs volley fired. At this time none of the twenty-fives had opened up."

Pollard: "My orders were to move out on line, but in the heat of battle the formation ended up more in the form of a wedge moving southeast to support the recon unit. While this move was underway, the LAV-ATs updated the enemy count to more than 75, with about half being tanks."[40]

The two forces flailed at each other in the darkness. First Platoon was engaging targets to 2nd Platoon's front. Pollard's XO radioed "They're shooting over friendly head. Someone from 1st Platoon says they're shooting too close, so I told them to cease fire and then shift to the southwest and get on 2nd Platoon's right flank." This shifted the platoon from the left flank to the far right.[41]

The Iraqis fired tank guns and Sagger missiles, and ". . . the CO just missed getting hit himself with a Sagger."[42] Pollard ordered the LAVs to back out of main gun range.

One of the LAV-ATs had misidentified its target: it was Green Two, another LAV-AT. Kendall: "This AT (Green Two) got off its round. As it was tracking [the flight of its missile] an AT back behind it—how it got behind it I'm still not sure—fired. It came through the bottom right troop hatch on this one, and hit all the other missiles I guess." (Wreckage indicated it was hit on the left hatch).[43]

The vehicle disintegrated in a huge fireball. "There were no secondary explosions. Nothing. This whole thing just went up." With radio contact lost, some believed it had been hit by Iraqi fire, others hoped that the radio had merely failed. Observing the friendly fire incident, the Recon platoon mounted its vehicles and evacuated the OP. The LAV-25s began to fire ineffectively on the Iraqi tanks. To their rear, Alpha Company was moving forward. Charlie Company 3rd Tank was alerted, then ordered to stand down.

Kendall continued to move forward to aid the Recon team. "At this time the ATs had all pulled back."

Control of air assets passed to D Company. Pollard: "On the suggestion of one of the platoon sergeants, a LAV-25 would fire 25mm high explosive rounds, and the LAV-AT s would guide them on target using their thermal sights. Once the vehicle was on target, a section or the entire platoon of LAV-25s would fire on the designated target. This resulted in hundreds of little explosions showing up on a group of vehicles, which the aircraft could then find and target."[44] TOW vehicles detected Iraqi tanks trying to flank the company, and destroyed three of them as Pollard backed up even further. At this point the LAVs were, by Pollard's account, five or more kilometers from the enemy at OP-4.

An OV-10 dropped an illumination flare which fell 25m from Red Two, an LAV-25. At 2300hours two Air Force A-10s rolled in on the ground battle. A radio operator tried unsuccessfully to deflect the A-10s. Pollard: "One of the scouts in back of the vehicle jumps out to try and bury it [the flare]. None of us knows what it is. The FAC tells the XO that's a friendly mark. . . . Just about that time, it all seems like it happened instantaneously, Red Two explodes. . . . That was the first time I got scared. . . ." A Maverick missile ripped the turret off the vehicle, killing seven Marines; survivors were the infantryman who had dismounted, and the driver who was blown out of his hatch.

"I talked to my TOWs and my platoon down there, and they said no, we haven't been flanked. Next thing I thought, my initial thing to do, was I wanted to get the hell out of the area, but I didn't know where to go. . . ."

About 2305hours Pollard called Alpha Company, who verified that he had not been struck by their fire. "The XO tells me he thinks it might have been friendly air." Pollard ordered his company to back up another kilometer.

By 2351hours Delta withdrew behind Alpha Company's screen line. Bravo Company also moved in on Alpha's south [right] flank, where the attached LAV-ATs requested permission to fire on observed vehicles. Captain Hammond ordered the TOW gunners to dismount and shoot a compass azimuth to determine the true location of the targets. The simple exercise revealed that the TOW unit was disoriented; the strange vehicles were friendly LAVs. Hammond froze LAV-AT fire until dawn.[45]

To the north near OP5 Charlie Company with a section of LAV-ATs and a section of LAV-M mortar carriers had fired on Iraqi positions earlier in the evening, and pulled back. At 2230 they were ordered to take up blocking positions around OP5, then ordered north to repulse Iraqi units that had crossed the berm east of OP6. The LAV-ATs fired at Iraqi vehicles, which fell back, leaving their infantry to secure OP6.

Around dawn Greg Michaels's section of LAVs from Alpha Company nervously entered al-Zabr to locate missing Recon men. They also recovered the driver of the vehicle struck by the Maverick, suffering from burns and concussion.[46]

At 0720 Alpha Company called in air strikes, and Captain Shupp instructed the aircraft to positively identify and count all his vehicles before any attacks. When the Cobras took ground fire from a building, Michaels fired on it but his chain gun malfunctioned and could fire only AP rounds. Alpha screened the area while battered Delta Company and graves registration personnel recovered the dead. Maintenance personnel repaired Michaels' ammunition feed mechanism, happily informing him that the replacement part was stripped from the wrecked vehicle.

Two LAV-MEWSS (Mobile Electronic Warfare Support Systems) appeared and located the enemy radio jammer that had caused such mischief in a building on the Kuwaiti side; a TOW missile destroyed the building and jammer.

Sergeant Hernandez was assigned to capture the crew in an enemy tank parked alongside another building on the Kuwaiti side. His scouts rounded up four prisoners, who turned out not to be the tank crew. They had fled the fighting around OP6 to the north, only to wander into the middle of the OP4 fight. Alpha then retreated to leave Delta in possession of the OP4 area.[47]

All these attacks were merely distractions from the main assault, where *15th Mechanized Brigade* pushed along the coastal highway. At 2230hours a strong force of tanks, armored cars, and personnel carriers loaded with infantry entered al-Khafji.[48] The Iraqis secured the city at about 0200. Marine observation teams pulled back, but two reconnaissance teams remained inside the city. A third powerful Iraqi attack farther inland was bloodily repulsed.

About 26 kilometers to the rear at 3rd Tank "We were in position, ready to go counterattack, should they be successful and start pushing south," said Mummey. "We were at 100 percent alert, and 'I wonder how long they think we can do this?'"

The counterattack of 29 January–1 February was primarily a Saudi and Qatari effort, with air and artillery support from the Marines. The Saudi-Qatari force destroyed ninety armored vehicles, killed thirty enemy soldiers, and took over five hundred prisoners. Marines from 3/3 staged an unsuccessful effort to recover two American soldiers captured when they blundered into the occupied city.[49]

The performance of the LAV battalions reflected the doctrinal ambiguity that still plagued these units. They performed well as a screening force, and checked the advance of a major attack, but commanders were unprepared to exploit their success. The LAV crewmen had also gained confidence from prisoners who revealed that although the 25mm could not penetrate tank armor, the explosions overwhelmed the sighting mechanisms, blinding the tank.[50]

The Marines had assumed too much of the enemy. In the aftermath Mummey and other NCOs were sent up to see the wreckage. "We started looking at the equipment. 'These guys are raggedy. These are just front line guys, and maybe the guys behind them are better. Maybe these guys are just fodder.'

"We could tell their equipment could still shoot, move, and communicate, but we knew they weren't Marines or Americans."

FINAL PREPARATIONS

Airpower advocates were still confident the Iraqis could be pounded into submission with no need for a costly ground campaign. Trapped in the desert, the front-line troops grew increasingly cynical. Intelligence assured them that the 105mm main gun of the M60 tanks could not penetrate the frontal armor of the T-72, though Army experts at Aberdeen Proving Ground knew better.[51]

After the al-Khafji affair, the forward logistical bases seemed less vulnerable. In retrospect Dennis Beal thought "That worked okay. Certainly with the Iraqis, because they couldn't do anything. They weren't gonna come out and take the initiative to get any of this stuff. You could have put it a hundred meters in front of the objective and they wouldn't have gone to touch it. Ninety-eight percent of these people were conscripts. They didn't want to be there. There's no leadership below about the battalion level at all. No one takes any initiative."

In the long deployment the old M60s were rapidly wearing out. Rick Mancini's tank developed a major fuel line leak, with little hope of getting a replacement. Maintenance Chief Randy Crane improvised. "So he gets a soda can, cuts it in half, slaps some putty on there, bubble-gum, on the pin leak, puts that thing on there. . . . That held the thing together as far as letting the fuel go through. Had no problem whatsoever. It was stuff like that he was just invaluable for."

Senior tank officers like Buster Diggs won an unseen battle. The Marines faced a mechanized force, and the shock power of large tank formations might be the key to survival. "The primary driver for that was a retired Lieutenant General by the name of Marty Steele. He started talking about tanks as a maneuver element." A new generation of tank officers instituted the practice of integrating whole companies of tanks with infantry battalions to form "tank heavy" and "mech [infantry] heavy" teams. "We would tell people 'No, you don't get a tank platoon, you get a tank company. If you want tanks, they come in companies.'

"He [Steele] kind of pushed for the tank battalion as a kind of maneuver element."

In 3rd Tank "At Twentynine Palms we always gave up tank companies which always led the infantry's attacks."

Diggs was impressed by "The reasonableness of people to listen to you. Infantry officers. To say that 'I've done a battalion CAX.' I would tell people 'This is how we do this,' and everybody would listen."

Al-Khafji had enormous impact on the campaign that was to follow. Observing Iraqi timidity in the attack and poor tactical coordination and control convinced Marine commanders that the enemy would not be as resolute in the defense as originally feared. The performance of the Qatari and Saudi units in Al-Khafji provided leverage to demand a bigger role.

The original plan was for 1st Marine Division to breach the defenses and 2nd Marine Division to exploit the breach. Rehearsals demonstrated the potential for a massive traffic snarl. Colonel Joe Fulford called Diggs in one day. "One day he said 'Buster, I just want to get everybody out there. I want to see what all these vehicles look like. We just got out in the field so we could *see* all this stuff.'

"But I kept emphasizing to him . . . I'm not fighting sixty tanks and twenty-five mechanized vehicles and TOWs. I'm just fighting four company commanders. I said to him 'You're just fighting four battalion commanders, boss. We'll do most of the work amongst ourselves.'"

More confident, Army General Norman Schwarzkopf allowed the Marines more leeway in the planning of their assault. The Marines entered the campaign with limited breaching capabilities, so breaching equipment was scavenged from all over the globe. Two weeks before the offensive began MGen Bill Keys, commanding the 2nd Marine Division, convinced LtGen Walter Boomer that the 2nd Marine Division could effect its own breaches. The plan changed again.[52] The Marines would smash through along a broad front. Arab contingents would follow, and veer east to attack directly toward Kuwait City. Other Arab forces would advance to the west of the Marines, and then veer east to join in the assault on the city.

The greatest fear was that Iraqi artillery fire would rain down upon the engineers as they cut breaches through the obstacles.

The two Marine divisions would use fundamentally different approaches to breach the defenses. On the 1st Marine Division front, east of The Elbow, infantry of Task Forces TARO and GRIZZLY would infiltrate through the first minefields by stealth, and screen the mechanized breaching forces, Task Forces RIPPER and PAPA BEAR. The LAI units would be among the first through the breaches, to delay counterattacks.

Third Tank was designated to join Task Force RIPPER, built around Jim Mattis's 1/7, with a "tank heavy" team under Diggs and a "mech heavy" team under Mattis. Diggs: "I gave him my best company commander [Rick

Mancini], for two reasons. A, he's gonna be on my right shoulder, leading Mattis's attack, and B, even though Mattis is a smart man—obviously, he retired a four-star—I knew he didn't quite understand this, he just had to follow Rick. Plus I could talk to Rick; we shared all the frequencies, I could talk to him if I needed to.

"He [Mattis] gave me his best company commander, because I don't know nothing about trench clearing. I don't want to know nothing about trench clearing. If we needed to clear trenches, you just tell [Bob] Hathaway 'Hathaway [Alpha/1/7], you need to clear that trench.' My captains were better than I was. My XO was far better than I was, Mike O'Neal, far better than I was."

On the 2nd Marine Division front north of The Elbow, artillery and aircraft would screen while engineers boldly cut six lanes. The assault battalions of 6th Marines would lead the breaching effort while the attached Army Tiger Brigade and the 8th Marines would move through to exploit the breaches. The 2nd Marine Division would be the most powerful Marine division ever fielded, with 257 tanks.[53]

Gonsalves's Third ("Blue") Platoon of Charlie Company, 3rd Tank got all the "toys," first a blade mounted to the platoon sergeant's tank. "Then we got two mine plows. . . . We got these towed assault bridges, I think we got them from the Israelis. They were these long bridges. We got two of them. At first we didn't know what to do with them. Do you tow them, or do you push them? We experimented to figure out how we're gonna move these things." Fortunately Sergeant White was "Like a MacGyver. He figured out all these neat ways to detach the bridge from inside the tank with a rope and clip. It was really interesting.

"Then one day they showed up with this thing called the Roller-Dude. It was a roller, so they said 'Okay, give it to Gonsalves's platoon and let them figure out how to use it.' You would plow, push the mines out, and push the roller through the minefield to blow up any [remaining] mines."

Eventually two tanks in each platoon were fitted with special equipment. A tank fitted with a rake or plow and an AAV-7E1, the engineer variant of the personnel carrier, would work in tandem. The tank would approach the minefield and halt, and the AAV would fire a Mine Clearing Line Charge (MICLIC) over the tank.

Tanks fitted with rakes, plows, or rollers would clear mines not detonated by the MICLIC. "Every tank but mine was either pushing or pulling something," said Diggs.

Rick Mancini's company was "Training, and training, and training. . . . You can't just do this thing once or twice. You have t do it continuously." Mancini decided to run one exercise at the dark of the moon. "Partway through I said 'Stop! Don't nobody move.'" At dawn "You look around and you see one hell of a gaggle of tanks pointing every which a way. . . . I said 'Well, this is good learning point for all of us. We got a lot of work cut out for us.'"

Iraqi infantry was dug into positions with inadequate overhead protection from air and artillery fire. There were no alternate firing positions once a defensive line was penetrated. Tanks were dug in to anchor infantry positions, with little or no provision for mobility. The most effective aspect was camouflage, as windblown sand obscured many positions, and Marines reported that enemy infantry would appear and disappear, seemingly from nowhere.[54]

James Gonsalves was a bit dismayed by what little he knew of the plan. While Army units with the most modern equipment made a sweeping movement through empty desert, "We were going straight up the middle. There was no maneuver, straight up the gut here. . . . We got the tanks with the least amount of armor, we can't see at night, we don't have the latest in range-finding gear. It was kind of like 'This is crazy!'"

The Iraqi Army was heavily mechanized. The vast majority of tanks were modeernized T-54/T-55s, Type-68 (Chinese) equivalents, and a few heavier T-62s. The more modern T-72 made up about ten per cent of the force.[55] The most potentially effective arm was their artillery.

The buildup continued until the very last minute, and Alpha Company 4th Tank was among the last to arrive. During a quick exercise in Oman, there were more problems. Crabtree: "The fording kit restricts the exhaust. One of those tanks [crew] didn't listen, and ended up blowing the engine because they didn't take off the fording kit, or they were just hauling ass." Once in Saudi "We took them off the night before the ground war started." In line for ammo issue, "We're listening to the BBC saying the ground war has started." One tank had to be left in Kuwait with a broken drive train part.

OPENING MOVES: 17–23 FEBRUARY

Civilians have the idea—reinforced by the news media—that major military operations can be conducted as an instantaneous "bolt from the blue." In reality, even "surprise" operations like the final assault into Kuwait require weeks of preparation, and once begun the momentum is inexorable. Coalition forces began to shuffle from defensive to offensive postures, leaving some sectors lightly defended. One solution was the creation of Task Force TROY, a bogus force to assume the place of 2nd Marine Division.

Lt. Freitus and his logistical coordinator, Master Sgt Graham had signed for ". . . an incredible amount of gear, munitions, ordnance, all very expensive electronic wizardry." Now the logistical bean counters insisted upon an itemized accounting. After sleepless nights, the two finally decided that the imminent fighting was more important. They had not yet discovered that expensive and "important" items could simply disappear without a trace in the overall wastage of war.[56]

On 18 February reconnaissance elements filtering into Kuwait discovered that the defenses had been poorly maintained: wire was sagging and broken, mines exhumed by the wind. The Marines marched north under a soaking rain that turned the desert to miserable sludge.

At noon on Thursday 21 February 2nd LAI Battalion began crossing into Kuwait, screening the preparations of the 2nd Marine Division, searching for alternate breach points for the Tiger Brigade. The Iraqis responded with artillery and mortar fire, and armored counterattacks. In the running battles Iraqi artillery was deluged with air strikes and counter-battery fire. The armored reconnaissance units expanded their activities to the east along the Marine front.

Mummey was surprised that after the rains "The desert had a velvety look to it. It had a green fuzz growing on it because of all the moisture. It looked kind of unreal." British tank transporters moved them forward. "We start creeping closer and closer [to the border]." They found that ". . . every time we'd move, there would already be holes dug for us. They had engineer units going ahead of us. Holes where we'd pull the tank in there, and you couldn't see out of the hole unless you were standing on top of the turret."

Just after midnight on 22 February GRIZZLY moved across the border and halted just south of the mine barrier. Iraqi artillery was silenced by Marine counter-battery fire, but GRIZZLY withdrew in the face of Iraqi tanks. That evening TARO began its muddy slog into Kuwait, and settled into positions just south of the mine barrier.

The infantry of GRIZZLY were planning to breach the mine barrier when Iraqi deserters appeared along undiscovered paths through the mine-fields. Marines rushed down to meet them, had the Iraqis mark the return path with chemical lights, and pushed a company through to the other side. GRIZZLY had its breach. On 23 February the task force commander decided on his own authority to push a battalion through the gap as a "reconnaissance in force."

GRIZZLY pushed on, hampered only by Iraqi defectors heading south, and stumbled across an unexpected mine belt. It had to be breached by probing for the mines with bayonets in the rain, mud, and darkness.

Behind them other units were moving into place. Despite advanced reconnaissance by senior NCOs, there were problems caused by communications breakdowns and malfunction of the Polaris satellite navigation system; the Marines fell back upon civilian GPS. On a tank in Freitus's First Platoon the camouflage net came adrift and tangled in the drive sprocket. It was laboriously cut away, the others jettisoned.[57]

On the 2nd Division front units pushed up to the berm in the blackness. One of B/4th Tank's new vehicles, WHEN'S CHOW? (Sgt. John Gibbert) was left behind because of a fuel leak and joined the company only fifteen minutes before the assault.[58] The war had not yet begun but some units were up to twelve miles (20km) inside Kuwait.

NOTES

1 Simmons, *Getting Marines to the Gulf*, p. 3.

2 Except as otherwise noted the account of this unit's arrival is excerpted from Freitus and Freitus, *Dial 911 Marines*, p. 43–47.

3 Moore, *Out Front At the Front: Marines Brace for Task of Clearing Mines*, p. 186.

4 Blount Island is a civilian contract facility near Jacksonville, Florida.

5 Estes, Interview with Dennis Beal, 2009

6 Michaels, *Tip of the Spear*, p. 22–23.

7 Freitus and Freitus, *Dial 911 Marines*, p. 48–49.

8 Estes, *Marines Under Armor*, p. 185.

9 Freitus and Freitus, *Dial 911 Marines*, p. 80–8.

10 Estes, *Marines Under Armor*, p. 185.

11 Freitus and Freitus, *Dial 911 Marines*, p. 90–92.

12 Anonymous, *This Was No Drill*, p. 25.

13 Ibid, p. 30.

14 Michaels, *Tip of the Spear*, p. 34.

15 Dacus, *Bravo Company Goes To War*, p. 9 and Zumwalt, *Tanks! Tanks! Direct Front!*, p. 74.

16 Ibid, p. 74.

17 Michaels, *Tip of the Spear*, p. 6–8.

18 Battlefield Assessment Team, Armor/Antiarmor Team, *Armor/Antiarmor Operatons in South-West Asia*, p. 15–16.

19 Ibid, p. 15, 28.

20 Ibid, p. 10–11.

21 Michaels, *Tip of the Spear*, p. 81–83.

22 Ibid, p. 102–105. Michaels does not provide first names for most of the persons mentioned, presumably to preserve privacy. Here this practice is followed wherever accounts are drawn from his book.

23 Freitus and Freitus, *Dial 911 Marines*, p. 99–100.

24 Michaels, *Tip of the Spear*, p. 70.

25 Ibid, p. 77–78.

26 Freitus and Freitus, *Dial 911 Marines*, p. 80–84.

27 Moore, *Storming the Desert with the Generals*, p. 108.

28 Michaels, *Tip of the Spear*, p. 35–36.

29 Freitus and Freitus, *Dial 911 Marines*, p. 72–73.

30 Anonymous, *5th MEB Deployment to SWA*, unpaginated; Anonymous, *The 1st Marine Division in the Attack*, p. 145.

31 For example, Michaels, *Tip of the Spear*, p. 102–103.

32 The following account is from Michaels, *Tip of the Spear*, p. 117–125.

33 Winicki, *The Marine Combined Arms Raid*, p. 54–55.

34 The following account is from Michaels, *Tip of the Spear*, p. 126–129.

35 The observation posts were at first numbered in the order in which they were established. The Marines tried to re-designate them in geographic order from the cost inland, and the dual numbering systems cause considerable confusion; see Wintermeyer, *The Battle of al-Khafji*, p.9 for explanation.

36 Wintermeyer, *The Battle of al-Khafji*, p. 13.

37 Anonymous, *A War of Logistics*, p. 157.

38 Wintermeyer, *The Battle of al-Khafji*, p. 15, Michaels, *Tip of the Spear*, p. 134.

39 There are discrepancies in the timing of events; times reported here are taken from Cureton, *Interview with Captain Roger L. Pollard*. Subsequent quotes from Pollard are from this interview. There are also discrepancies between published accounts and Cureton, *Interview with Lieutenant David Kendall*.

40 Pollard, *The Battle for OP-4: Start of the Ground War*, p. 48–49.

41 Cureton, *Interview with Captain Roger L. Pollard.*

42 The company XO speaking in background of tape in Cureton, *Interview with Captain Roger L. Pollard.*

43 Cureton, *Interview with Lieutenant David Kendall.*

44 Pollard, *The Battle for OP-4: Start of the Ground War*, p. 49.

45 Michaels, *Tip of the Spear*, p.141.

46 Ibid, p.142–144.

47 Ibid, p. 145–154.

48 Wintermeyer, *The Battle of al-Khafji*, p.22.

49 bid, p. 27–28.

50 Estes, *Marines Under Armor*, p. 137–138, Estes, *Learning Lessons From The Gulf?*, p. 93, Michaels, *Tip of the Spear,* p.156.

51 Freitus and Freitus, *Dial 911 Marines*, p. 127; see also Gilbert and Swan, *T-72 In Iraqi Service*, unpaginated.

52 Anonymous, *Rolling With the 2d Marine Division*, p. 149; Anonymous, *Special Trust and Confidence Among the Trail Breakers*, p. 89.

53 Anonymous, *Rolling With the 2d Marine Division*, p. 150–151.

54 Battlefield Assessment Team, Armor/Antiarmor Team, *Armor/Antiarmor Operatons in South-West Asia*, p. 4–5, 9.

55 Ibid, p. 4–5, 9.

56 Freitus and Freitus, *Dial 911 Marines*, p. 141.

57 Ibid, p. 149, 153–154.

58 Dacus, *Bravo Company Goes To War*, p. 11.

CHAPTER THREE

The Storm Breaks

*One may know how to gain a victory, and know
not how to use it.*—Pedro Calderon de la Barca

DAY ONE: SUNDAY, 24 FEBRUARY

A T 0400HOURS 1ST Marine Division units moved across the border and into Kuwait. Their primary objective was the big airbase at al-Jaber.

At 0620 RIPPER encountered the first minefield, and paused to bring up Buster Diggs's Team Tank and the breaching equipment. Diggs: "General Boomer, when he showed up for the breaching operation, pulled me aside and said 'You guys really do this, huh?' He's the commander! I go 'Yeah, we really do.' This is our time; it's our turn. It's open desert. This is a tank-artillery battle.'"

"When you do the breach," explained Mummey, "you've gotta have a support force, so we would pull up and be the support force engaging targets in and around the breach site." Once the enemy was suppressed, "One-seven Marines with Alpha Tanks attached and their engineer det(achment), they would come up and actually do the breaches. Then we would turn into the assault force, and push through those lanes. . . ."

The infantry was still unaccustomed to working with tanks. An AAV stopped near Mummey's tank and disgorged its load. "The grunts pulling up in front of me and setting up mortars in front of my tank while I'm trying to shoot at stuff on the other side. . . .

"These kids start setting up their sixty millimeter mortars right in front of my tank. I'm trying to yell at them 'Hey! Get the f out of the way! We're shootin' here!" The grunts were heedless of the danger. "I told my wingman we need to move to the right. . . . because the grunts weren't gonna move."

The engineer breach teams included marking teams mounted in AAVs, MICLICs on AAVs, larger launchers on trailers, and tanks with plows or rakes to proof the lanes. Lieutenant Kurt Kempster of the engineers attached to Task Force PAPA BEAR explained that the seven-man marking teams were to ". . . mark an entrance and an all clear point on the mined lane, and if we were to do the breach at night they would also mark the left-hand side of the lane as we plowed through it with chem lights. . . ."[1] The first charges would be fired from trailers towed by tanks. "The gunner and tank commander are tankers, and the driver and the loader are engineers."

In Mancini's second breach the lead AAV with the MICLIC got stuck in a crater. "I had to bring up the second one that we had. At the same time I brought up my M88. . . . And he was doing recovery ops while under mortar fire. It took twenty-four minutes on the second breach. . . . instead of the twelve and a half that it did on the first one."

The MICLIC line charges arced through the sky. "These things are de-signed not to go off until they're completely deployed. This one detonated about twenty feet in front of the tank, in mid-air, so it was pretty exciting. I'm sure they got a real good ride. It was pretty exciting for me, being out of the hatch about a hundred feet from them." The MICLIC failure rate was high, and engineers had to go into the minefields on foot to detonate them.[2]

Staff Sergeant Louder, an engineer in Task Force PAPA BEAR: "We went ahead and shot a charge out . . . over a tank. That charge went out no problem. Once we shot our second charge out, it didn't go off. We decided to throw another charge out to detonate the first charge. That charge didn't go off." Louder and LCpl Householder tried to set it off manually.

Traffic was beginning to back up. "By this time it was pitch black, we're at MOPP level four, communications was next to nil; everybody and his brother was on the net. . . . We came down to a plan of actually going out-side and blowing this. . . ." The problem was how to reach a safe distance before the huge charge exploded.

Householder worked out a plan. "The tank swung his turret around so that the hatch would be on the close side. He plowed out about five feet from the charge so that the lane was pretty clear. We had pre-made blocks of C-4 already. Ran down into the minefield about fifty meters, put the block of C-4 on top of the line charge, and ran back into the tank and waited for it to go off." Householder felt "Like a white Carl Lewis. It was kind of scary, but I knew it had to get done so there's no way around it."

In the awkward MOPP suit and gas mask, he stumbled over the tank's mine plow ". . . and ended up slipping and sliding up the front of it in these rubber boots. I kind of for a minute thought I wasn't gonna make it," but had seconds to spare. With the tank hatch open ". . . dirt came flying in. Kind of felt like the tank moved, popped up kind of like a front stand or a wheelie."

In Task Force RIPPER's area one of Gonsalves's tank commanders, Sergeant White, was awarded a Navy Achievement Medal for dismounting to help detonate a failed MICLIC. By 0645 Lane Four was through the mine barrier.

At RIPPER's Lane Three, MICLICs failed to detonate. Even when everything goes perfectly the refracted blast wave from the ground creates a "skip zone," and fails to detonate mines. Tanks equipped with mine plows and "roller dudes" moved in to proof the lanes.[3]

The plows dug into the earth, pushing sand and rock—and live mines— to either side. At Lane Three Sgt Scott Helm's tank detonated a large mine, blowing off the left front road wheel and bending the suspension arm. Mummey: "As tanks and vehicles would pass, the weight and the vibration would make those things come back down and an Alpha Company tank ended up hitting a mine. That started clogging up the lanes; it started making for a bad deal because nobody [could] get through the lanes. . . .

"That's when Staff Sergeant Johnny Cruz got his Bronze Star, for going in there and pulling that tank out of the breach. He ground-guided an M88 up there, through the minefield. There's still mines laying around."

Kessinger thought that "Breaching the mine fields for real was easier than what we made it out ourselves in training. You're supposed to do it that way. In training you always had this feeling like 'you gotta get through those lanes as fast as you can; we're gonna be taking artillery and direct fire the whole time.' And we're thinking twenty-five per cent of us aren't gonna make it through the first minefield." Once the drill commenced, "I drank a Dr. Pepper going through the first minefield belt. I saved my last Dr. Pepper. I was saving it, [but] I figured 'Well, this is it. I'm going to at least have something good to drink.'"

There was ". . . no direct fire that I knew of and no appreciable artillery. Every once in a while a round would land somewhere. You could hear it. But nothing even close to doing any damage."

Once through the breaches, the tank companies deployed into company wedges. The company command group was free to roam, supporting all three platoons.

RIPPER's tanks came under artillery fire in the zone between the first and second mine belts. Diggs: "He [General Trainor] told us about the artillery and how they'll build barriers in front of you and then the artillery will shell you. Call me cocky, but there's no contest between attacking tanks and artillery. There's just no contest. We shoot too fast and too accurately. You might get one or two of me, but you're gonna die."

Mummey: "Every time we'd start getting hit with that, we'd have to hold up, start buttoning up. You'd lose your situational awareness. Then you gotta call and make sure it's not our stuff coming down."

Tim Frank was a TOW gunner assigned to Task Force RIPPER. "We did take a multiple rocket attack that first night. We had a bunch of people get hurt. A couple of vehicles got hit directly. But that didn't really change anybody's attitude. It just fired us up a little bit. . . . With our technology and our training and our technology, we had this sense that we're invincible."[4]

As C Company sat stationary in the gap, artillery ranging rounds began to walk toward the tanks. Freitus intuited that the Iraqi observer was in a concrete water tower just out of range of the tank guns. He had the FAC riding in place of his loader, Captain Jeff Butler, "paint" the tower with the MULE. After several Hellfire missiles struck the tower, the artillery became lighter and more erratic. Others were less fortunate. When Lt. Leo Teddeo moved forward and began heaving smoke grenades to obscure the retreat of the rest of his company, he was wounded in the arm by shrapnel.[5]

Iraqi infantry proved less resolute. Freitus: "Heads bobbed up and down, looking much like a prairie dog town." A lone Iraqi ran from position to position. When the Marines did not kill him, the desert began to sprout improvised white flags. [6]

Kessinger: "I've got the weapon (an M16A3) shouldered as we're driving through this tight area that has a lot of these bivouacs everywhere. And I know that if someone had jumped out of one of those bivouacs, I wouldn't have been looking for a white piece of paper that they were waving over their head. I know I would have just pulled the trigger. Thank goodness nobody jumped out."

On the far side of the minefield the tanks spotted the first T-55 and T-62 tanks, dug in up to their turrets with aprons of building tiles, gravel, or canvas tarps in front of them.

Fatigue and distraction were already beginning to tell. Kessinger's tank came around a knoll: "I saw the T-55 about nine hundred meters to my front, and I (said) 'Gunner! Sabot! Tank!'

"The colonel wasn't doing anything. I just looked over. 'Colonel! Load a sabot!'

"He just jumped down and threw one in the breech. It's like 'Oh shit! *I'm* the loader!' He had a few other things on his mind I suppose.

"I fired and missed, and whatever tank was on my right blew it away. I'm not even sure I missed. You know sabot rounds go right through those things, but I didn't see a flash. But within a few seconds it was gone." An Alpha Company tank to the right fired a HEAT round "So it makes a big boom."

This was one of only two rounds the command tank fired as the crews quickly realized that most of the Iraqi tanks were abandoned. Targets were prioritized, and enemy tanks were low on the list. The other round was fired at a radar-controlled ZSU 23-4. "That's a priority target. Any time that pops up, you gotta take it out, no matter whether it's manned or not because of the threat to aircraft."

The battalion wheeled toward its first objective, taking Iraqi positions in the flank. In Mike Mummey's platoon the platoon leader and platoon sergeant's tanks carried HEAT loaded and the two wingmen carried sabot. "If a target popped up, whoever had the right ammo took the shot." Mummey spotted a tank positioned to fire into the hull of any American tank coming over a ridge. "There's this T-55 sitting there in his dugout, his little hull-down, and he's aiming up at the ridge to his right front, which would be my far left. He thought that's where everybody's coming from. All of a sudden here comes Charlie Company sweeping around, and you see the turret came over to us, and went back up the hill, and started to come back to us.

"About that time [to] my gunner, [Corporal] Gilger, I said 'HEAT! Tank!' and he let fly a round. Hit it right at the turret ring, and knocked the shit out of it. . . . That was about a nine-hundred meter shot.

"Gray smoke and dust came out of every aperture on that tank; every bolt hole. It just started burning."

The tankers had never fired depleted uranium (DU) ammunition into a hard target. "The first time 'Fritz' Fitzpatrick, my wingman, shot a tank, I spotted and I was like 'I can't tell!' It looked like it hit, but it didn't make a flash. . . . It just went in and out, and I saw this big sand plume go up behind it.

"I told him 'Lost! Reengage!' His gunner—about the time he was getting ready to shoot [again], all of a sudden right around the turret ring it erupted with fire."

On the 2nd Marine Division's sector to the west, the assault began at 0430hours with a deluge of MLRS rockets on the Iraqi artillery. At 0530 the first waves of Marines crossed the berm, and by 0600hours reached the first mine belt. Task Force BREACH ALPHA was lavishly equipped with engineer equipment.[7]

At 0700hours 2nd LAI moved through the breaches. Without artillery and mortar support, C Company withdrew in the face of enemy fire, but was back through the berm by 1030hours. Behind the mine barriers were Iraqi infantry and tanks in revetments; the LAV-25s used cannon fire to mark them for the TOW vehicles.[8]

The thirteen M1A1 tanks of B Company, 4th Tank, operated in support of 1/8. Two tanks from each platoon were fitted with mine plows to "proof" the lanes. Breaching began at 0600, and operations at four of the lanes went like clockwork. At Green 5 and Green 6, the two lanes on the right flank, there was no such good fortune.[9]

At Green 5 the first engineer AAV fired its second MICLIC over an electrical power line, bringing down the lines. The plow tank crept underneath, trying to work a path through the unscathed minefield. Two huge blasts lifted the tank into the air and dropped it onto its broken tracks. The trapped AAV laboriously backed out of the lane.

A second M60 plow tank carefully worked its way alongside the crippled tank, followed by the AAV. The AAV fired its third and last MICLIC, but when the M60 moved forward to proof the lane, it struck another mine, blowing off a track. There were no casualties, but two of the precious plows were lost.

A third plow tank towing a trailer-mounted MICLIC picked its way into the lane and fired the MICLIC, started a new lane segment, and successfully proofed it. Another engineer AAV moved up and fired its line charge, but when the plow moved in to proof the lane it struck another mine. The AAV backed out along the long, crooked lane. The company blade tank moved in to push the disabled tanks aside, but could not budge them. With three tanks disabled and one trapped in the crooked, dangerous lane, further work ceased.

At Green 6 the commander of the lead M60A1 plow tank, SSgt B. M. Shaw, was startled when the engineer AAV following him hit a mine. Shaw had missed the marker stake, and was trapped inside the minefield. The second vehicle team moved up 70 meters to the left of Shaw's tank. The engineers' MICLIC failed to detonate. An engineer picked his way into the minefield, manually primed the line charge, and ran back, racing the thirty-second fuze. A second MICLIC failed to detonate. Another engineer went out and primed it, diving to safety in the AAV just as the 1,800-pound (818kg) charge detonated. Lane Green 6 was finally cleared.

Chip Parkison's B Company, 4th Tank was supposed to be first through Green 6, but he did not trust the lane. He ordered Viper 2, Sgt Robert Trainor's M1A1 with a plow, into the lane. Re-plowing the lane was dangerous; it deepened the track ruts, allowing the tank to belly out on the center. Two thirds of the way through the lane a huge blast lifted Viper 2, and radio communication was lost.

While Parkison was trying to raise Trainor on the radio, Trainor's Platoon Leader, Warrant Officer Larry "Wolf" Fritts, radioed that he was going into the lane. Ominously, undiscovered anti-personnel mines were detonating under Viper 1's tracks.

Inside the disabled tank a fine spray from a ruptured hydraulic line was mistaken for the oily spray from a chemical mine. The turret crew donned protective masks and climbed out, but Lance Corporal Arnel Narvaez was missing, trapped in the driver's compartment by the turret overhang.

Fritts shouted to Trainor to manually traverse the turret, an agonizingly slow process. When Trainor opened the hatch, Narvaez looked up, unscathed. Fritts crammed Trainor's crew into his own vehicle.

Pressure from the rear was growing; columns waited behind the unbreached minefield, exposed to potential artillery or chemical attack. By 0724hours the Marines were pouring through the other breaches. On the division's exposed left flank 2nd LAI Battalion spread out to screen the advance, and the division forged ahead, its right flank exposed.

Shaw, sitting in his trapped tank, could take no more. He inched his tank forward, plowing through the minefield to connect with the cleared section of Green 6. Incredibly, Shaw reached the far side, then turned and re-plowed the lane.

Parkison's tanks flooded through Shaw's path. Captain Brian Winter, Parkison's Executive Officer in Predator 5, struck three mines. The first two were anti-personnel mines, but the third was a shaped-charge anti-tank mine. Shaw's plow had flipped it upside down, and the deadly explosive jet blew harmlessly into the sand. After hours of delay, the Marines flooded through Green 6. Shaw was promoted to Gunnery Sergeant and awarded the Silver Star.

Past the minefield the tanks encountered Iraqi T-55s, BMP-1s, and infantry that chose to fight. LCpl Lance Miler, in an M60A1, destroyed a BMP-1 at 3,200 yards (about 2,900m), the longest known kill. He was awarded the Bronze Star.[10] After fifteen minutes of concentrated fire from tanks, artillery, and infantry the enemy began to surrender in droves, waving anything white.

The problems at Green 5 and 6 caused momentum of the attack to shift left, so C/2nd LAI shifted left to give A Company maneuver space. Sergeant Richard Smith's LAV-25, serial number 521666, had seemed a jinxed vehicle so the crew arranged a blessing by a bemused Catholic chaplain. Six-six-six encountered a low ridge at right angles to the main defensive line. Smith barged over the ridge and into a hornet's nest of enemy infantry. At nightfall the LAVs halted, but all night the Marines called in supporting artillery and engaged entrenched infantry with 25mm fire.[11]

Delays in breaching the mine barrier led to confusion. An artillery battalion was to cover the breaching operation, with priority of fire controlled by 3rd Tank. When breaching operations began, priority switched to Task Force RIPPER, and 3rd Tank's missions slid down the "hopper," or queue. Diggs: "If there are any missions left they don't go away. They just shoot Mattis's missions first, then when I became the exploitation force, then I had priority of fire [again]. So that's what the artillery did; started shooting those missions.

"All of a sudden here comes Dual-Purpose ICM down on me, but I see it's coming from the back! I wounded somebody. Stupid battalion commander." Diggs accepted responsibility for the error. "A lot of people should have known to do that, but mainly the battalion commander."

Task Force SHEPHERD's movement was impeded by surrendering Iraqis, which Greg Michaels likened to "a steady, moving roadblock." The LAI was tasked with escorting prisoners to the rear, but they were simply overwhelmed by numbers.

The LAI had to struggle through the congestion to assume the lead. SHEPHERD's task was to screen against forces in the al-Wafrah and al-Burqan oil fields while the division seized the al-Jaber airfield.

Tanks were ". . . dropping plows because a plow's pretty heavy. Dropping plows, dropping bridges" according to Diggs. He was concerned about prisoners because "I had also heard that was a tactic the Iraqis used on the Iranians. They'd have whole units surrender, and then the Iranians would be bogged down processing prisoners. I have one infantry company, and we had pistols on the tanks. That's not gonna work. I'm not gonna give up my infantry company, [because] now I'm bogged down. Luckily there was a police station we could see . . . and I got my boss and said 'Let's send them over there, and let's just keep moving.' I was afraid they were building more obstacles."

The advancing Marines encountered clouds of toxic hydrogen sulfide gas leaking from broken wellheads and pipelines. Hydrogen sulfide in small concentrations smells like rotten eggs. In higher concentrations it kills before a human can register the odor. Raised pipelines provided other impediments for the scouting LAVs. All this took place under a drizzle of oil raining down from the burning wells of the al-Burqan oil field. Visibility dropped to ten feet (3m).

Sam Crabtree awoke in his Humvee and thought, "Oh look, the sun's coming up. Then I look at my watch, and it's three in the morning. . . . We were seeing the oil fires." Another night ". . . It was raining oil. Not like a full-on rain, but sprinkling oil from the sky." Mancini also remembered the oil fields "You go into them it's just black. . . . At the other end you got soot all over you."

Kessinger:" We had to navigate to the left of the oil field. In the oilfields with the pipelines, you can get caught up in one. It's like a spider web. You really have to get outside of one and move completely around it. If you go

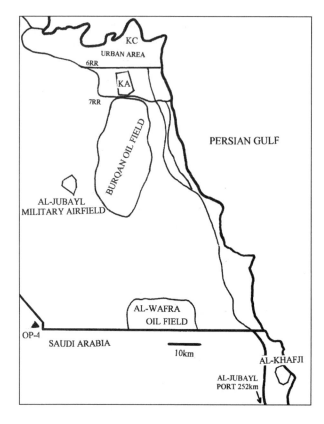

MAP 1. Marine Corps Area of Operations, Kuwait. KA=Kuwait International Airport; KC=Kuwait City; 6RR=Sixth Ring Road; 7RR=Seventh Ring Road.

inside of it you get channeled and locked in, and there's no way out."

Before the battle the T-72 seemed to have the advantage. "It outgunned us," said Diggs. "My concern was could I penetrate it. But once we get tacked into oil smoke, our engagements were one-five-hundred (meters) and below. Well, that one-oh-five would go through anything. . . . In that oil smoke we could kill anything at one-five-hundred or less."

In the darkness and confusion Sergeant Anecito Hernandez's LAV engaged two trucks at point-blank range. Tracers fired from a vehicle behind arced past Hernandez's head, and later he found bullet holes in ration boxes stacked behind his head. At nightfall the company set into positions overlooking a broad depression, using the thermal sights of the LAV-ATs to control mortar fire.[12]

The confusion was such that one tank commander reported sighting a T-72, but was so excited that the neglected to fire on it. A crewman in one of the tanks farther back in the column spotted the turret slowly traversing, and fired on the enemy tank.[13]

By 1445hours the M1A1s of Bravo Company, 4th Tank moved through the second mine belt, after providing long-range covering fire for the breaching team. At about 1600hours the advancing Marines routed Iraqi infantry, and Bravo Company prepared to engage supporting Iraqi armor.

The ensuing Battle of the Candy Canes was a long-range gunnery duel in which the Iraqi armor stood no chance. Drawn up in a line atop a ridge in advance of the main force, B Company could see Iraqi armor dug into revetments and in motion beyond a row of red and white electrical transmission towers that gave the battlefield its name. Beginning at about 1650hours Marine gunners systematically picked off vehicles at extreme ranges; Sgt Glen Carter's STEPCHILD destroyed an Iraqi T-55 at 3,750meters (4,100 yards). When the company withdrew, they had destroyed ten tanks, a ZSU-23-4 self-propelled gun, and numerous trucks and light vehicles.[14] At 0230hours the tanks went into a defensive coil and the tankers caught a few hours sleep. A kilometer to the northeast a paved road ran along a raised roadbed through a complex of seemingly abandoned trenches and bunkers; the elevation was enough to mask the ground behind the roadbed.[15]

At about 1730 on the 1st Marine Division front a pair of Iraqi T-62s tried to engage Task Force RIPPER, and one was destroyed by TOW missiles fired by the infantry's CAAT (Combined Anti-Armor Team). Freitus watched in amazement as an artillery round landed between two Humvees. The TOW gunner on one never took his eyes off the target. "After seeing that I stood in the open hatch of my tank and pretended to be brave."[16] The tanks and CAAT began to systematically destroy Iraqi tanks in their revetments.

During the final movement to the al-Jaber Airfield, Freitus had the "most personal moment of the war . . . taking a human life." Inflicting death for a tanker is usually an impersonal thing. In the moving tank Freitus shouted and the driver, Florence, veered to one side to avoid a shallow pit. Glancing to the side, Freitus saw a concealed mortar pit. Unable to fire down into the pit, he grabbed a grenade and heaved it, ducking down to avoid the fragmentation.

There was no explosion. There was no pin in his hand. He jumped off the tank, ran back, and lobbed a second grenade. As he turned to run back to the tank, he glimpsed motion in the pit. Leaping onto the fender in a single adrenaline-fueled bound, he dove into the hatch as the grenade exploded, touching off a series of secondary explosions.[17]

Kessinger: "We would get mortar rounds landing on us, and then three minutes later come across a mortar crew. You bastards! You were just shooting at us a second ago, and now you're giving up!"

Kessinger said that "I never saw one cruel thing done to anybody in the enemy [forces]. If anything I saw compassion and kindness. Anybody I know is pretty proud of that. When we took any prisoners, we would give them our food. They desperately needed it, that's for sure. At the same time they were the enemy and you're just not gonna trust them."

Early nightfall in the 1st Marine Division zone found Task Forces TARO and GRIZZLY screening the flanks near the breach sites, PAPA BEAR had veered east. Task Force RIPPER sat observing the al-Jaber airfield, and SHEPHERD the al-Burqan oil field. B Company of SHEPHERD had been detailed to provide security for the division's Forward CP. Losses were unexpectedly light; only one Marine killed in each division.

The Iraqis spent the night struggling to mount armored counterstrikes. Conditions were so bad that when Freitus dismounted from his tank to relieve himself, he became disoriented. This was extremely dangerous; a man on foot can easily be run down if the tanks are suddenly called upon to move. He found the tank by feeling for the tracks in the sand, and following them until he smacked his head into the tank. He climbed back aboard, too embarrassed to tell the others what had happened. Saint Elmo's fire, a blue-green electrical charge, began to dance on the tall antennae, eventually bathing entire tanks in its eerie light.[18]

About 2300hours Iraqi artillery impacted near Richard Smith's 666; he ordered the driver to move backward about 75 meters, but the next volley impacted just to the right of the vehicle, shredding tires, the exposed exhaust system, and external gear. The blast dislocated Smith's right shoulder; he popped it back into place himself. There was no pain, but several minutes later Smith felt something running down the arm. Blood.

Smith told his gunner, "Vega, I've been hit."

"Bullshit! You have not" Vega replied, before seeing the blood.

Keying his radio, Smith called out "Fat Chick! Fat Chick! This is Juggernaut. I've been hit," and passed out. Hustled to the ambulance and then to an evacuation hospital, Smith's last sight of Kuwait was towering plumes of smoke.[19]

The company Headquarters Platoon from Alpha Company, 4th Tank was an assortment of wheeled vehicles and an M88A1. Crabtree was part of the Contact Team, with another mechanic and two communications men riding in a Humvee. If there was a problem, "We would drive over there first, and see if we could fix it. That way we didn't have to drive the Eighty-Eight all over the place." The mechanics worked at night within sight and sound of the enemy. "We had red-lens flashlights, but sometimes you had to use a white light because you just can't see. Either you do it, and take the chance, or try to throw a poncho over you or something. . . .

"One night we had to pull the pack because the starter was out. We were taking it off the CO's vehicle . . . because he was going to be riding a Humvee. We were set up in a coil, and then a Cobra flew in . . . and started firing that Gatling gun they have." Crabtree was later told someone had spotted an RPG team, but he never determined what had happened.

DAY TWO: MONDAY, 25 FEBRUARY

Iraqi efforts to mount a counterstrike were hamstrung by other uncertainties. An amphibious feint by 4th MEB pinned Iraqi mobile forces in place. Air strikes pounded Iraqi mechanized forces. The midnight counterattack foundered in the face of fire from TOW units and LAVs.

Kessinger: "We were shaving . . . so our gas masks would fit properly. Otherwise you might not really care in the middle of combat. We were taking turns, keeping two guys on the tank while two guys got cleaned up, and then switching." Kessinger and his driver (Kevin Moroni) went first.

"Colonel Diggs and Corporal Galvan jumped down, and no sooner did they get basically undressed than 'Gas! Gas! Gas!' came over the radio, the battalion tac. That's your worst nightmare, that you get hit by something like that."

The crewmen were still in the MOPP suits. "Colonel Diggs has his gas mask on, and all of us, but Corporal Galvan had left his gas mask in the gunner's position. He literally dove in. I'm sitting in my suicide chair [a small folding seat] in the cupola. I don't know how he did it. He dove in right through my lap head first down into the loader's seat. His feet were still sticking up in my lap, and he's fishing around for his gas mask, trying to hold his breath this whole time.

"I remember him looking up at me and saying 'Isn't life a bitch?'"

About 0530hours at C/4th Tank's coil position south of al-Abdaliyah, Captain Alan Hart walked over to Parkison's tank to discuss the morning plan. Corporal Brad Briscoe from Hart's 2nd Platoon ran to notify them that he had detected vehicles moving across the company front. Hart assured him it was just "our own amtracs running around in front of us again."[20]

Briscoe told Hart's gunner to turn on his thermal imaging system and watch their front. The distinctive sounds of tracked vehicles was now audible: Parkison told Hart "Those aren't our amtracs."

Hart ran back to his tank, shouting for everyone to man their positions. Diving into his own tank, he looked through the thermal imager. Hart could see at least a dozen T-72s emerging from behind the raised road, moving across the company front from left to right.

An Iraqi tank unit was stumbling blindly, but not one of the Marine tanks was prepared to seize the advantage. It requires about three minutes to bring the tank's systems online. Crewmen were still on the ground; firing the big cannons could injure or kill them. Hart shouted "Tanks! Tanks! Direct front!" Then over the radio, "Predator 6. This is Hawk 1. Enemy tanks to my direct front! Come on line *now!*"

Hart shifted his tank to a safe firing position. His gunner, Cpl Lee Fowble, calmly said, "Sir, we've got to shoot! They're traversing!" Fowble had ordered the loader to chamber a round, saving precious seconds. Hart gave the order to fire, and the lead T-72 exploded in a fireball.

In the T-72 ammunition is stored throughout the crew space. The ammunition is consumable case; the casing of the propellant charge is combustible. Virtually any penetration of the armor sends hot shards and molten droplets of the tank's own armor spraying throughout the interior, igniting the stored propellant. The usual effect was to blow the turret about ten meters into the air.[21]

Fowble shifted his aim and fired, destroying another T-72. Another M1A1 came on line, but two rounds from the other tank flew high. Fowble destroyed another T-72.

Hart realized that three parallel columns were emerging from the blind ground, crossing the roadbed at separate points. Interspersed were Chinese-made armored personnel carriers and wheeled vehicles. This was an Iraqi tank battalion.

Titan 4 [SSgt Jeffrey Daucus] came on line and engaged the enemy; 1st and 3rd Platoons were moving into firing positions on Hart's flanks. Humvees from 1/8 began to move up and engage with TOW missiles.

From his position behind Hart's 2nd Platoon, Parkison watched the T-72s ineffectively return fire. After each round the autoloader in the T-72 elevated the gun tube as it ejected the stub of the projectile case, loaded a new round, and in theory returned to the point of aim. In fact it seldom did. The rate of fire was slower than with a human loader, and the gunner could not use the impact of the missed shot to adjust the aiming point.[22]

Hawk 4 [GySgt Alfonso Pineda] had a malfunctioning laser range finder, but the problem was resolved in time to fire the last round of the seven-minute battle. By 0700 the surviving Iraqis were surrendering. In what became known as the "Reveille Battle" the tanks and TOW vehicles had destroyed thirty T-72s, four T-55s, and seven APCs of the veteran *8th Mechanized Brigade, 3rd Armored Division*.[23]

Dazed and wounded Iraqis staggered and crawled about among burning tanks. Some dragged or supported wounded comrades. Cautiously the Marines and Navy corpsmen began to gather in the wounded and dying to offer what help they could to the men they had been slaughtering.

Chip Parkison watched as a haggard Iraqi officer, desperately clutching a jacket to his chest, surrendered. A childhood friend had been killed in the tank behind his, and the jacket from the body of his "brother" was the only memento he could take home to his friend's parents.

An even bigger Iraqi force of two brigades had concentrated in the al-Burqan oil field, on Task Force PAPA BEAR's right flank, on the extreme right. At 0815hours a massive artillery bombardment flushed tanks out into the sights of attack helicopters, TOW gunners, and the M60A1s of 1st Tank Battalion. The shooting dragged on into the afternoon.

About 13km (8miles) to the north the 1st Marine Division's Forward Command Post was situated west of the Emir's Farm, an oasis right out of an old film. The FCP was screened by a rifle platoon and seven LAVs from B Company of Task Force SHEPHERD. Just before 0930 enemy artillery began falling around the FCP.

The command LAV-25 with Captain Hammond and his gunner, Sergeant Randy Buntin, was on the extreme left of the line. A Humvee came racing over a low ridge about a kilometer to the north, the driver shouting, "There are vehicles coming this way."[24]

Little help was available because of the fight around the oil field. Buntin opened fire as the first BMP topped the ridge, but at close range the rounds flew over the low-slung BMP.[25] Buntin fired again and the vehicle staggered to a stop as the 25mm rounds ripped through the thin frontal armor. Infantry spilled out the rear hatches, but Buntin's coaxial machine gun jammed. The Iraqis fired an RPG, and Buntin raked them with HE from the 25mm.[26]

A second BMP was stopped by 25mm fire, but the infantry fired another RPG that detonated in the trees above Buntin's vehicle, spraying the scouts in back with splinters. When a third BMP emerged, Buntin stopped it and raked the emerging infantry with HE. The LAVs counterattacked, driving the Iraqis back toward the Emir's Farm.

Two helicopter gunships came to the aid of the FCP, their attacks directed by 25mm fire. Around the LAVs lay 38 wrecked Iraqi armored vehicles. Captain Hammond was awarded the Navy Cross, and Buntin the Navy Commendation Medal with V.

Iraqi resistance was erratic. When A Company resumed its advance they blundered into the flank of an enemy tank battalion. The LAVs destroyed a dozen tanks before the enemy began to surrender. The dismounts shoved thermite grenades in to the gun tubes, and the company moved on.

When the company encountered another entrenched unit a BMP started out of its revetment. Buntin's gunner fired but missed.[27] Gunner Uke adjusted the range. The BMP lurched and rolled back into the pit. Uke had killed the enemy vehicle at an incredible range: 3,500 meters.[28]

To the west, Task Force RIPPER was engaged in clearing bunkers that dotted the al-Jaber Airfield. At 1400 GRIZZLY was called forward to assist.

The Iraqi utilization of "environmental warfare" was highly visible and to some extent effective. Vast pools of crude oil impeded mobility and channeled attacks. A vast pall of oily smoke rendered night vision devices useless. Close air support was restricted by smoke that hung just above the ground.[29]

As the 2nd Marine Division with the attached Tiger Brigade neared Kuwait International Airport, an Iraqi armored force emerged to fight. The "steel rain" of the division artillery smashed the attack. Bravo Company, 4th Tank picked off single enemy vehicles in a running gun battle as the Iraqis fled.

At day's end the 2nd Marine Division had destroyed 248 tanks and captured 4,500 prisoners, but were ahead of the 1st Marine Division, with their right flank exposed. On the far right the JFCN force had made even slower progress. The Marines had also bypassed the large oil fields, and no one knew what lurked inside those infernos.

The Marines were well ahead of schedule, throwing the overall plan into disarray. Schwarzkopf started the advance of V Corps across the desert fifteen hours ahead of schedule, rather than halt the Marine advance.

Buster Diggs: "By the second day I could see that we were going to win this. Now I didn't want anybody leaving this war as a bad feeling. I told the boys 'I got plenty of ammunition. If there's a doubt, shoot. There's no question—shoot. But if you think somebody's gonna surrender, give them the benefit of the doubt. I don't want you to leave here thinking you're a butcher. I don't want you to leave here thinking you butchered human beings.'"

DAY THREE: TUESDAY, 26 FEBRUARY

Radio Baghdad announced at 0135 that Iraqi forces had been ordered to withdraw from Kuwait. The time for any sort of orderly withdrawal was long past: communications were smashed, and Coalition forces relentless. For the Iraqis it was every man for himself.

West of the city B/4th Tank was positioned near a road intersection when Iraqi forces began their breakout attempt. The tankers slaughtered the Iraqis, destroying nine T-62s, ten BMPs, three BTRs, an MT-LB tracked carrier, and trucks.[30]

Alpha Company of TF SHEPHERD was positioned north of the al-Burqan oil field, facing northeast while the attached LAV-ATs faced back into the gloom of the oil field. The heat overwhelmed the thermal sights, and at 0400 a MT-LB trundled undetected right into the Marine lines. Fortunately it was filled with deserters and loot. When the company resumed its advance, it was quickly overwhelmed with prisoners. Finally an exasperated artillery observer, Corporal Reese, commandeered an abandoned bus and drove a load of prisoners to the rear. In retrospect Michaels thought this took considerable courage, given the enthusiasm with which Marines and aircraft were shooting up "enemy" vehicles.[31]

Allied commanders decided to block the retreat by bombing the Mutlaa

Pass, where the main highway passed through a low ridge. Aircraft pounced on the huge snarl of vehicles, and vehicles took to the open desert, only to encounter their own minefields. West of the city other forces tried to flee, only to be smashed by Marine tank and infantry fire.[32]

Task Force GRIZZLY veered west to complete clearing the al-Jaber area, and close up with 2nd Marine Division. TARO would lag behind to clear the Al-Burqan oil field. The bulk of the Marine forces headed for Kuwait City, impeded by hordes of surrendering Iraqis.

The advance began at about 0700, hampered by a morning sandstorm. The tank battalions utilized differing methods to deal with the dozens of Iraqi tanks and armored vehicles that littered the landscape. Some tickled each enemy vehicle with machine gun fire, and then blasted any that responded. Others simply shot up anything that looked alive.

By 1130 hours 3rd Tank was drawn up along a phase line marked by a line of electrical power pylons. Iraqi artillery began to lob single rounds at the stationary tanks. The rounds walked closer and closer, as Freitus repeatedly called in a "Snowstorm," the signal for an artillery attack.

When battalion was slow to respond Captain Ed Dunlap, the CO of Charlie Company/3rd Tank, took it upon his own authority to move past the phase line. Freitus, on the extreme right, was the last tank out. He ducked into the cupola, and a good-luck medallion on a chain around his neck snagged inside the cupola. As he tried to free the chain and stick his head back out, the tank lurched violently sideways, and then moved on. Freitus freed the chain and looked back as explosions deluged the vacated position. Shrapnel had shredded the right-side sponson boxes, the gear in the gypsy rack, and severed cables on the exterior of the turret. Eerily, Freitus's grandfather, an Army tanker, had been spared when the same medallion became entangled during the Battle of the Bulge in 1944.[33]

Third Tank moved past The Orchard, a tree farm south of the Seventh Ring Road encircling the city. A single road channeled the tanks. Gonsalves was worried that it was the perfect ambush site. "Just a really bad situation. . . . It was pretty much the middle of the day, and you couldn't see fifteen feet in front of you. It was pitch black." The lead elements stumbled upon two Iraqi armored vehicles in the gloom. The tanks engaged them at point-blank range, and the major danger was from the explosion of ammunition in the enemy vehicles.[34]

Charlie Company moved up to help infantry clear The Orchard. Kessinger: "From a Marines' perspective, it looked like just beautiful fields of fire. If they had set up any kind of direct fire down those [rows of trees], it was perfectly flat ground. You couldn't tell from one row of trees to the next what you were coming up on, or what you were going around."

On the outskirts of The Orchard, Mummey "Sat there and looked off to the east, and all of a sudden 'BOOM!' I see this fireball come flying across the ground. I knew what it was right away: that's a Sagger. I reached down. . . and I popped my smoke." Freitus heard Sergeant "Fritz" Fitzpatrick radio "Sagger! Sagger," the code word for an anti-tank missile launch.

Mummey: "In the confusion "A bunch of people saw that smoke and thought it was a chemical attack. All these tanks were stopping and guys were throwing on their masks.

"My wingman, Sergeant Fitzpatrick, he was on my right front right next to this berm, and right about the time he dipped down, that missile hit right at the top of the berm and exploded and the missile bucket went flying over his back deck." Fitzpatrick began to back up in an effort to throw off the missile launcher's aim.

"I started shooting HEAT and Wooly Pete at these shapes that were over there; I didn't know exactly what was over there. Then all of a sudden 'BOOM!' Here comes another fireball. It starts to go behind me, and it's going toward Lieutenant Freitus's tank." Freitus could clearly see the fuzzy fireball of the second oncoming rocket's propulsion system, and the puffs of smoke that marked the operator's course corrections. The tanks were now taking heavy machine gun fire from the rear.[35]

Mummey: "He [Freitus] starts doing a backward stagger-dance, left-right, hanging on throttle, sharp turns." Now two tanks were backing up and a potential danger was collision with other tanks still advancing. "I remember watching that missile pass behind my tank. It just nosed in the dirt in front of his tank, just a few meters. About that time we had this kid—I forget his name—but he was the FO on Captain Dunlap's tank. I was like, 'We need immediate smoke on the right flank.'

"All of a sudden there's this big curtain of smoke from all these Wooly Pete rounds hitting. We're sitting there hung up. There's this big agricultural orchard, so we're all cramped up. We got people shooting at us from the orchard, we been taking these missiles from the right flank. Colonel Diggs decides to let One-Seven out to go clear the woods. They go! These grunts been in the amtracs for three days, now all of a sudden they weren't coming back, man.

"I remember he said on the radio 'I've let loose the dogs of war and I can't get them back!'"

Kessinger: "That's when Colonel Diggs dismounted his infantry. . . . They set up out in front, and you could hear the rat-a-tat-tat of machine gun fire, all the things the infantry does. Colonel Diggs had his head down and was shaking his head back and forth. He looked up at me and said, 'What have I done?' He let the crunchies loose."[36]

Contrary to Diggs's fears, Kessinger said the infantry ". . . patrolled the whole area. It's what they do, and they did it very well. Made sure we were safe. Then they mounted back up and we moved on through. There was really no other way the colonel could have gotten it done. A tank couldn't do it, and it had to be done on foot. But I know he felt like he lost a little control when he let all those guys loose."

The tanks were still taking small arms and machine gun fire, so Dunlap ordered the tanks forward again. Mummey: "My gunner, Gilger, once again he pumped some HEAT rounds. He makes 'em stop what they're doing. Lieutenant Croteau [1st Lt. Christopher Croteau, the leader of 1st Platoon, Charlie Company] takes the lead. He punches through on the side of this agricultural thing. The grunts are over there on the left. They're going through so we've got to watch out. Up ahead of us we didn't know there were all these people heading toward Iraq. . . ."

When the lieutenant saw all the traffic on the Seventh Ring Road, "He starts lighting them up. I'm the fourth tank back; I'm pulling drag. I don't know what the hell's going on. Nobody does.

"Then we pushed up there and continued shooting. Eventually most of them jumped out of their vehicles or they just kept on driving. It was getting dark. . . ."

The advance of the tanks was blocked by another obstacle belt. "We had all these (burning) vehicles, and they're starting to cook off and every-thing. . . . We're within a hundred meters of them, so we like need to get out of there. But we had the rest of the battalion coming up behind us. So Scott Martin in Third Platoon, he had a plow on his tank. . . . He came up and pushed through that wire so we could get to the other side of the freeway, which we didn't know was also an Iraqi position, but they had already hauled ass out of there. . . ."

When Mummey drove past Martin's tank he saw that ". . . [barbed wire] was caught up in his mine plow, caught up in his sprockets. . . . He was standing there—he didn't have his helmet on—he was just sitting there looking at his sprockets all wadded up with wire, looking all bummed out."

To the west the 4th Marine Division and Tiger Brigade were closing off the Iraqi escape route, destroying the odd Iraqi vehicle: a half-dozen tanks, a BMP, more MT-LBs and trucks for Bravo/4th Tank.[37]

At 1400hours the 1st Marine Division began the final advance on Kuwait International Airport, with Task Forces RIPPER, PAPA BEAR, and SHEPHERD arrayed from west to east. Infantry forces would seize the airport proper, while SHEPHERD enveloped it from the east. During the night the infantry-heavy Task Force TARO moved north to the outskirts of the airport and prepared to finalize the capture.

The Marine units began to arrive at final phase lines late in the day, securing the southern outskirts of Kuwait City and a major road junction northeast of the city. Arab forces would enter the city proper, and premature darkness was approaching. General Walter Boomer ordered the 2nd Marine Division to reorient and sweep west to cut off any remaining Iraqi forces.

Near nightfall Buster Diggs's tanks sighted a strongpoint about 8,000 meters out. The Iraqis sent an emissary to negotiate surrender. "He came to us and said, 'Listen, we're honorable men. We're not gonna just surrender to you. You've got to send one tank out there tomorrow morning and shoot at us, and then we'll surrender.'

"I'm going 'Jeez, if I send a tank out there, only one tank's gonna go, it's got to be mine.' And my tank commander's gonna go 'You gotta be shittin' me Rommel! I'm not goin' out there!'

"So we started out the attack with Hellfire missiles downrange—we had tactical helicopters with us all the time. So we lased, and then they surrendered.

"Those people would have done that; they would have done exactly what they said. But some Iraqi mothers didn't get their sons [back]. You know, it just didn't sound right to me. Plus, remember, I had never been in combat. I didn't really know how the game was played, except you're on this side of the gun tube put your hands up or you die. That's all I could do."

At 1500hours Alpha Company of TF SHEPHERD received orders to resume its northward advance. Just outside the International Airport the leading vehicles encountered two armed pickup trucks that sped across the front of the column. Gunners on multiple vehicles began to blaze away at the trucks, sending tracers on all sides of the leading LAVs until Sergeant Ramirez quelled the firing by shouting over the radio. The incident resulted in a thorough chewing out for gunners and vehicle commanders.

The LAVs halted in defensive positions at a complex freeway intersection west of the city. The one-way lanes divided by concrete barriers formed a maze through which the vehicles could not find a clear path. Efforts to breach the barriers failed, and the position remained insecure through the night.[38]

Marine reconnaissance units were steadily filtering through the collapsing Iraqi units all along the front. A recon team contacted Kuwaiti resistance fighters who led them into the city to reoccupy the American Embassy. This reportedly enraged General Norman Schwarzkopf, who had organized a showpiece operation to liberate the Embassy.[39]

To 3rd Tank's front, they could hear the engine idling in an abandoned tank. Kessinger: "The Colonel called Recon and wanted . . . them to go out and turn it off so we would know if it was a manned vehicle or not. The Recon lieutenant said 'I don't know how to turn a tank [off]!'"

Several of the tankers had experience with Soviet vehicles. Corporal Mullendore volunteered to go. "I turned and looked at him, and he looked at me, and I said 'Did you just volunteer?'"

The sheepish Mullendore replied, "'That was the last thing my Dad said before I left; don't volunteer for anything.'"

Kessinger: "We're in the back of this Humvee, driving out toward this engine noise, and they had a TOW vehicle with them. Somebody thought the vehicle started moving, and they just lit it up with the TOW."

DAY FOUR: WEDNESDAY, 27 FEBRUARY

Kuwait International Airport is a sprawling complex with a huge multi-story passenger terminal. About 0330hours the LAV companies breached the perimeter fence. Scrambling to move through the narrow gaps, vehicles collided with each other until crews placed chemical light sticks in the side windows. The only delay was when Alpha Company encountered mines along the western perimeter.

Infantry from TARO and the LAVs from SHEPHERD moved out across the grounds. Nervous crews fired through some of the huge terminal windows, mistaking reflections of burning oil wells for movement inside. The LAI scouts had to search the sprawling terminal building, and the airport was declared secure at 0900hours.

To the north, the Tiger Brigade was the first ground unit to encounter the carnage along the "Highway of Death." Coalition aircraft had destroyed some of the leading vehicles, and pounced on others trapped in the jam.

Army Special Forces and Navy SEALs arrived at the American Embassy amid a flurry of helicopters and video recorders. The Marine recon team was kept safely out of sight. The war was rapidly winding down. General Colin Powell was trying to rein in Schwarzkopf. Orders at last came down to cease offensive operations as of 0800, 28 February.

DAY FIVE: THURSDAY, 28 FEBRUARY

Arab Coalition forces passed through the Marine lines and into Kuwait City, to a joyous celebration—and vengeance on collaborators. The Marines had destroyed over a thousand tanks, as well as 600 other armored vehicles, and taken 22,308 prisoners. The cost was five killed and 48 wounded.

There remained a final, potentially bloody, "mopping up" phase. Fifth MEB received orders to clear Iraqis isolated and bypassed in the Al-Wafrah Forest along the Kuwait-Iraq border. Engineers and tanks had to breach another minefield to allow access. Sporadic and ineffective sniping delayed final clearing of the adjacent Al-Wafrah oil field until 3 March.

Diggs summarized the Marines' part in the operation. "I didn't know what Tenth Corps was going to do by the way. That was pretty compartmentalized information I did not know. I did not know they wanted us to get bogged down in Kuwait. That was the plan for us to get bogged down. . . . We just didn't cooperate."

Dennis Beal thought that "One of the reasons they got their asses kicked so bad was they fought the standard layered Soviet defense where all the decisions are made from a central command, and if you break the communications lines you couldn't talk to anybody up there and no one could get a decision. Because no one's gonna do anything on their own. My personal assessment of all of this was that up until this point in time I was always thoroughly impressed with the Israelis and how well they did.

. . . My final conclusion at the end of this thing was that the Israelis never were all that good. The Israelis were good, but they weren't as good as we thought they were, it's just that these people were so *woefully* bad. Shit, this is like an NFL team playing a high-school team. These people just didn't understand the operational art of war, they didn't understand cross-boundary fires, they didn't understand fire and maneuver. They didn't understand any of the things you need to understand to be successful in combat."

WINDING DOWN

For a brief while Marine units remained encamped inside Kuwait, sometimes under bizarre circumstances. B Company, 4th Tank moved south to a locale they called Pet Cemetery for the seventy or so carcasses of cows, horses, and—inexplicably—zebras. They remained amid the stench for ten days.[40]

One of the common observations was the extent and detail of the Iraqi looting. Early in the fighting Sam Crabtree's maintenance section had established itself in a small Kuwaiti town. "We set up in a defensive line, and to make sure that the buildings next to us were clear, we went to the houses. . . . All the electrical outlets had been removed by the Iraqis. There were vehicles scattered. I didn't see one of them that had four tires. They took anything and everything they could."

Though units were officially forbidden to fly American flags, Kuwaitis cheered the Marines' "unofficial" displays of American and Marine Corps colors. Third Tank remained in position for several days, then drove south through the city, "Where they were lining the streets, waving flags, and all the things you'd love to [see]," said Kessinger.

The 5th MEB re-embarked for a planned rescue mission in Addis Ababa, Ethiopia. The mission never materialized, and the MEB was soon dispatched to cyclone-ravaged Bangladesh.[41]

"We abandoned all of our equipment" explained Diggs, "because we knew we were going to transition to M1A1s so we didn't want the tanks. Then FSSG came in big time. We had ammo broken out, probably fired ten per cent of the rounds we had in the tanks."

Mike Mummey: "We were down there for four or five days. We had to take all the reactive armor off the tanks, had to lay all the ammo out on tarps, count every bullet, count every smoke grenade. Ammo techs would come by and classify it, then we had working parties taking it."

The M60A1s were left behind, but "Of course we're taking our pet tanks with us now. We got a pet T-72 and a pet BMP that we're taking with us. . . ."

Kessinger "Did an LTI inspection of the tank, signed off on the logbook, and walked away. That was a really strange feeling. You've just spent eight months in that tank, and you just turn and walk away. It was a little surreal, actually. You just leave a little piece of your life sitting back there." Mummey said, "I know my tank ended up being a reef off the coast of South Carolina. They demil(itariz)ed it at Albany (the Georgia Marine supply depot) and it got pushed off a barge off the coast of South Carolina. . . ."

For the Marines there were the same enemies that appear after every war: resurgent bureaucracy, REMFs, and former allies. Freitus was stopped at a Saudi National Guard roadblock, and an officer began to rant at Freitus about his many shortcomings as an infidel. Freitus snapped. He jammed his pistol under the man's chin and suggested that he and his men ground their prominently displayed weapons. Slowly backing away with the Saudi officer walking alongside, pistol under his chin, Freitus put some distance between the other guardsmen before speeding away. He expected repercussions, but none ever arose.[42]

Another incident could have come from any war. Freitus and others ventured into the mess hall at Camp 13—a paradise with flush toilets and PX.[43] A senior staff NCO "locked onto us like a Stinger missile. . . ." Berating the tankers for their ragged uniforms and slovenly appearance, and for carrying weapons in a rear area, he concluded that they were a disgrace to the Corps. Only midway through his tirade did he sense that he had perhaps gone too far, and backed off.[44]

AFTERMATH

The feared T-72 had proven to be a deathtrap. Lt. Brian Holmberg of Alpha Company, 3rd Tank described the effect of the old M60A1's 105mm: "One [sabot] round hit the frontal arc of the turret on a T-72. It went through the turret, engine, then out the ass-end."[45] Kessinger said that when a sabot round penetrates and fragments strike hydraulic lines and stored ammo. ". . . things start cooking off, and as they're cooking off there are little explosions. And then there's the big one that usually sends the turret thirty, forty feet in the air, tumbling like a hubcap. The thing weighs several tons, comes down, and it burns for a long time."

One Marine Corps study rather dispassionately reported that Iraqi tanks struck while in motion ". . . invariably had some amount of human remains contained within what was left of the crew compartment. In almost every case these remains were to be found in the driver's area up under the front glacis. Inspection also revealed dismembered body parts strewn about the vehicle and within 15–20 meters."[46]

In contrast, for Diggs it was "Nice to get everybody back, and not have to look at wives and stuff and say 'I'm sorry.' For a commander, that's just the greatest thing in the world.

Marines encountered the effects of the first "media war." Kessinger was at his mother's home. "Here I am at my welcome home party, but my step-father—who was a Vietnam veteran—had recorded on VHS tapes every single hour of CNN from the war. I had never seen it. Here I am wanting to watch this, while everybody else is having a party!" Kessinger thought, "It was kind of nice to find out what we were doing."

Life returned to garrison routine. Charlie Company 3rd Tank was quickly trained on the new M1A1. "We got used for every infantry officer course, dog and pony show, combined arms exercise," recounted Mummey.

Mancini thought that one of the legacies of the brief war was a better relationship between the infantry and tankers. Mattis requested that "his" tank company deploy back to the U.S. with the battalion. "That sent a very strong message through all the ranks, that the infantry wants to take care of us."

Short-lived Operation PROVIDE COMFORT was an effort to provide humanitarian assistance and stabilize the restive Kurdish regions of northern Iraq in the aftermath of the war. Isolated within the land-locked north, everything had to be ferried in by heavy-lift helicopters or fixed-wing aircraft, including a platoon of LAVs.[47]

There was considerable unhappiness about leaving Saddam Hussein in power. In the following years many—particularly the Kurds, who had openly sided with the Coalition—would pay the price. The Iraqi chemical arsenal remained intact. Iran, with enormous population, oil wealth, powerful military, and its unbending antipathy to the United States, was still a regional power. Saddam Hussein had once been a counterforce against Iranian ambitions, but the pet monster had grown beyond control to itself become a threat. The monster was temporarily prostrate, but all things considered, it was thought best that Saddam remain in power. But like some monster in a low-budget movie, he would rise again.

NOTES

1 Cureton, *Interview with members of OCD attached to Combat Engineer Detachment, Task Force Papa Bear*. All quotes attributed to engineers involved in the breaching operations are from this group interview.

2 Cureton, *Interview with members of OCD attached to Combat Engineer Detachment, Task Force Papa Bear*, p.ii; and Bennett, *Battlefield Breaching: Doing the Job Right*, p. 19–20.

3 Bennett, *Battlefield Breaching: Doing the Job Right*, p. 19–21.

4 Fahey, *Interview with Major Tim Frank*.

5 Freitus and Freitus, *Dial 911 Marines*, p. 157–160; Lowry, *The Gulf War Chronicles*, p. 100.

6 Freitus and Freitus, *Dial 911 Marines*, p. 161–162.

7 Various sources cite differing times for the passage through the berm. Breaching assets included eighteen AAVs with roof-mounted MICLIC launchers, 22 AAVs to carry the engineer squads, 39 M59 MICLIC launcher trailers, and fifteen M9 Armored Combat Engineer (ACE) tractors. The supporting tank elements included two M60A1 blade tanks, four M60A1 with mine rakes, 16 M60A1 with mine plows, four M1A1 with mine plows, six M1A1 with mine plows, and four armored bridge layers (AVLB) built on the M60A1 tank chassis. Lowry, *The Gulf War Chronicles*, p. 101.

8 Michaels, *Tip of the Spear*, p.184–186.

9 The following account of 2nd Division and details of B/4th Tank's travails is primarily from *Braving the Breach* (sidebar) in Zumwalt, *Tanks! Tanks! Direct Front!*, p. 75–77.

10 Lowry, *The Gulf War Chronicles*, p.106.

11 Michaels, *Tip of the Spear*, p.186–188.

12 Ibid, p.192–202.

13 Moore, *Allies Used a Variation of Trojan Horse Ploy*, p. 103.

14 Dacus, *Bravo Company Goes To War*, p. 11–12.

15 Zumwalt, *Tanks! Tanks! Direct Front!*, p. 78.

16 Freitus and Freitus, *Dial 911 Marines*, p. 167.

17 Ibid,, p.172–175.

18 Ibid, p.193, 197–198.

19 Michaels, *Tip of the Spear*, p.188–190.

20 Unless otherwise noted the account of this action is drawn from Zumwalt, *Tanks! Tanks! Direct Front!*, p. 78–80.

21 Gilbert and Swan, *T-72 Tank in Iraqi Service*, not paginated.

22 Ibid.

23 This is the official count as listed in Zumwalt, *Tanks! Tanks! Direct Front!* There is some confusion over both types and numbers destroyed. Three T-55s (as shown

in Zumwalt's map), were dug into revetments and not part of the attacking unit. Some sources also include a single T-62 in the total. See for example *Aftermath of the Reveille Battle*, sidebar to Dacus, *Bravo Company Goes To War*, p. 14.

24 The account of the fight at the Emir's Farm is drawn from Michaels, *Tip of the Spear,* p.207–211. Note that there are serious discrepancies between this account and that provided by Lowry, *The Gulf War Chronicles*, p.129–130.

25 Battle sight is a quick-response sight setting that assumes a flat trajectory out to about 1700 meters. At half that range the very gentle arc of the rounds was carrying them slightly above the target.

26 Michaels, *Tip of the Spear,* p.209.

27 With the CO absent, at this point Buntin had apparently taken over the vehicle.

28 Ibid, p.212–214.

29 Battlefield Assessment Team, Armor/Antiarmor Team, *Armor/Antiarmor Operatons in South-West Asia*, p. 3.

30 Dacus, *Bravo Company Goes To War*, p. 14.

31 Michaels, *Tip of the Spear,* p.215–219.

32 Lowry, *The Gulf War Chronicles*, p. 140–141.

33 Freitus and Freitus, *Dial 911 Marines*, p. 208, 213.

34 Ibid, p.216–217.

35 Ibid, p.211–214.

36 Tankers often refer to infantry as crunchies—purportedly the sound they make when you run over them.

37 Dacus, *Bravo Company Goes To War*, p. 14.

38 Michaels, *Tip of the Spear,* p.221–238.

39 Lowry, *The Gulf War Chronicles*, p. 175.

40 Dacus, *Bravo Company Goes To War*, p. 15.

41 Anonymous, *5th MEB Deployment to SWA*, unpaginated.

42 Freitus and Freitus, *Dial 911 Marines,* p.244–246.

43 Camp 13, temporarily christened Camp Rohrbach by the US military, was a Saudi facility near the King Fahd Industrial Port built to house foreign workers. It was taken over by the military and housed among other units two Naval Construction Battalions (Seabees) renowned for providing their own creature comforts.

44 Freitus and Freitus, *Dial 911 Marines,* p.246–247.

45 As quoted in http://www.qrmapps.com/gw1/day1.htm

46 Battlefield Assessment Team, Armor/Antiarmor Team, *Armor/Antiarmor Operatons in South-West Asia*, p. 18.

47 Corwin, *BLT 2/8 Moves South*, p. 201.

CHAPTER FOUR

Intervallum

Those who have committed this evil act against the innocent, the women and children, to create thousands of widows and orphans, do not do so in the name of Islam. By the grace of Allah the Almighty, may justice be served upon them.
—Syed Abbas, supreme leader of northern
  Pakistan's Shia, 14 September 2001[1]

THE COMMITMENT TO the war in Kuwait had not absolved the Corps of its numerous other global responsibilities, only stretched it thinner. The burden of many of these missions would fall upon the new LAV units, simply because they were easier to transport and support in far-flung corners of the globe. In April 1991 Alpha Company, 3rd LAI Battalion was called upon for a relief operation following the eruption of Mount Pinatubo in the Philippines.

Mohammed himself had once issued a *haditha* proscribing Muslims from warring against Christian Ethiopians, who had once given refuge to his early adherents driven out of Arabia. Nevertheless kingdoms in what are now Ethiopia and Somalia descended into a prolonged period of reciprocal invasions and religious persecutions.[2] Later Britain and Fascist Italy warred over Somalia. In the aftermath of World War II no one quite knew what to do with it, so Italian administrators continued to govern until independence in 1960.

By the late 1960s Somalia had fallen increasingly under the influence of the Soviets, who abruptly betrayed Somalia, shifting aid to rival Ethiopia. In the aftermath the country spiraled into civil war between rival clans. Food was a weapon wielded by whoever controlled the ports. Famine claimed as many as 500,000 Somalis, and threatened many more lives. As part of a UN relief effort, the Marines were sent in to protect UN relief efforts.

The inclusion of armor in a humanitarian mission was the subject of considerable confusion. In the original plan the task force was to be built around the 7th Marines, and would include 3rd LAI Battalion (-), and the 3rd Amphibious Assault Vehicle Battalion (-).

At the last minute 1st Marine Division deleted a tank detachment from the operation's troop list, but Brig. Gen. Jack Klimp wanted the tanks. A reinforced platoon of tanks from C Company, 1st Tank was added, but at first the five tanks simply sat on the docks at Mogadishu.[3] The tank officer in charge, Mike Campbell, had only one NCO and four enlisted tankers, and shanghaied other Marines to fill out four tank crews; the fifth tank would be a spare. Unfortunately the 120mm ammunition had already been loaded back aboard ship and sent away by a conscientious supply officer who saw no need for it without tanks. The tankers were able to scrounge two boxes of .50caliber ammo, and 700 rounds of 7.62mm ammo from in-fantry and AAV units.

The missing ammo resulted in a bizarre "battle" when the four tanks were ordered to support a raid on a clan militia arms storage site mounted in response to the ambush of a UN convoy. Third LAI screened the sur-rounding area, and at 0647 hours on 7 January the Marine tanks, supported by TOW teams, barged into the compound. The tanks faced off with six Somali M47s. The Marines opened fire with machine guns; the Somalis thought the rounds were from ranging guns, and surrendered.

In another incident Marines heard engine and track noises in an area between them and an Army compound. Next morning there were three T-55s and two M47s in the field about 400 yards away. One M47 slowly tra-versed its turret toward the Marine tanks. Campbell ordered the crews to hold fire because he intuited that something did not seem quite right. Sud-denly a "crew" of children bailed out and ran. The Army had towed confis-cated tanks into the field as a dump site without informing the Marines, and the tanks had become toys.

For most of their tenure the Marines generally patrolled on foot, thereby gaining credibility in the local warrior culture, but the availability of the tanks was a reminder of the restrained power of the Marines. LtCol Malik of the Pakistani 1st Punjab Battalion said that "If the Marine tanks were not always with their infantry to at least intimidate and provide psychological support, but usually to draw away hostile fire, the Marine casualties would have been much higher." In a few cases tanks dispersed hostile crowds in a novel way, directing the 1,500 degree F (816C) exhaust of the turbine engine as a sort of "non-lethal" weapon.[4]

By March 1993 a food distribution network was functioning, and the Marines were replaced by American soldiers and UN troops of several nationalities. With the Marines went the heavy armor, and some analysts have suggested that the absence of a heavy armor reaction force, coupled with insensitivity to the nuances of the local "warrior culture," contributed to the disastrous Battle of Mogadishu depicted in the book and film *Black Hawk Down*.[5]

Well below the general public's threshold of perception, the volatile Middle East and Central Asia had again spawned another strange coalition of would-be tyrants who used religion to cloak ambition

Afghanistan was always more a state of mind than a nation. Since the time of Alexander the Great, centuries before the birth of Mohammed, the high, bleak deserts of the region created a fertile blending of cultures and a fiercely warlike people. The only unity between warring tribes and clans came when they briefly united in the face of an outside foe.

The Treaty of Gulistan that ended the First Russo-Persian War (1804–1813) precipitated a century-long strategic struggle for control of the crumbling Persian Empire and the independent Khanates of Central Asia. Imperial Russia sought to extend its boundaries south to secure a coveted warm-water port, while Britain was desperate to forestall any Russian influence in India. In 1838 Britain invaded Afghanistan and installed a puppet ruler, but by January 1842 the British were forced to retreat by unrelenting resistance. In the withdrawal over the snow-clogged passes, all but one Briton and a servant were killed; the two were allowed to live to carry the news of the disaster back to India.[6]

Britain again invaded Afghanistan in 1878, garrisoned strategic sites, and forced the Afghans to cede territory to British India (modern Pakistan). An uprising in Kabul in September 1879 killed the British Embassy staff and guards, and prompted another punitive expedition. Britain subdued the rebellion, forced the ruler Yaqub Khan to abdicate, and made his cousin Emir. Yet another revolt was put down by the British, who wearily decided to abandon Afghanistan to local rule, but retained control of Afghanistan's foreign policy in exchange for a monetary subsidy.

Alarm over the growing influence of Germany in Iraq finally brought about the 1907 Anglo-Russian Convention, which left Afghanistan in Britain's sphere of influence, and divided up moribund Persia (Iran).

The 1917 Bolshevik Revolution ushered in an era of Soviet expansionism, and the assassination of the Afghan ruler brought his son Amanullah to power. Amanullah immediately and rashly declared war on British India; wary of another entanglement, Britain conceded to let Afghanistan handle its own foreign policy. Amanullah played the British and Soviets against each other, trying to regain lost territory and establish a greater Pashtun tribal state. Amanullah also tried to modernize Afghanistan along the lines of Ataturk's secular Turkey, but in January 1929 was forced to abdicate by a fundamentalist Muslim revolt.

Amanullah's cousin, Mohammed Nadir Khan, crushed the revolt, and reigned as a conservative reformer until he too was assassinated in 1933. His son Mahammad Zahir Shah ascended the throne and continued many of the father's reforms under the tutelage of his uncle, Prime Minister Sardar Mohammad Hashim Khan.

The outbreak of World War II brought about another uneasy agreement between Britain and the Soviet Union, with Afghanistan coerced into expelling a large German expatriate community.

After Hashim Khan's successor was forced to resign, Zahir Shah instituted constitutional reforms which unfortunately paved the way for extremist parties, notably the Communist People's Democratic Party of Afghanistan (PDPA). In July 1973 former Prime Minister Mohammad Sardar Daoud Khan seized power in a military coup and abolished the monarchy, but his attempts at reform failed.

In April 1978 a PDPA coup overthrew Mohammad Daoud, murdered him and his family, and established a Communist "People Democratic Republic." The PDPA attempted liberalizing reforms, predictably resisted by religious conservatives in the small towns and countryside, leading to a simmering civil war.

With violence spiraling out of control, Soviet troops entered the country in December 1979. The US, operating in part through Pakistan's Directorate for Inter-Services Intelligence (ISI), backed a loose coalition of proxy resistance movements known as the *mujahadeen*, leading to ten years of brutal warfare.

A very minor player in the struggle was Osama bin Mohammed bin Awad bin Ladin, minor son of a prominent Saudi business family. The young bin Ladin's father divorced his mother, and he was raised by a step-father. Like many of the leaders of such movements, bin Ladin never completed his religious studies, instead developing his own even more extreme version of *Wahhabi*, an ultra-conservative and militant strand of Sunni Islam. Bin Ladin soon broke with the orthodox Muslim members of the resistance to found his own militant group, *al-Qaedah* ("the base") with a vision of extending *jihad* beyond Afghanistan and against the West in general.

The exhausted Soviets staged a much-publicized withdrawal from Afghanistan in February 1989, but continued to support the beleaguered central government. When the USSR collapsed, the Afghan government followed in April 1992. The many resistance factions fell to warring among themselves, and the Taliban—a coalition of fundamentalist scholars and former *mujahadeen*—ultimately seized control of most of the country. Dissidents were driven north to the Tajikistan border. Blocked by Tajik border guards, refugees were slaughtered by Taliban who conveniently ignored Islamic prohibitions against harming non-combatants. The Taliban imposed a repressive religious dictatorship over most of the nation. Oil money began to flow in increasing torrents to found *Wahhabi madrassas* (religious schools) throughout south and central Asia. These "schools" recruited students from among the poor, and many—though not all—taught only extremist *jihad*.[7]

Without a central government and with no law enforcement, remote, chaotic Afghanistan soon became a haven for various terrorist groups. Bin Ladin had been targeted for capture or death by Presidents Bill Clinton and George W. Bush; he and his followers had been harried from Bosnia and the Sudan. In collusion with the Taliban, *al-Qaedah* worked to consolidate its Afghanistan position. A necessary step was elimination of the Northern Alliance, a tribal-based resistance movement that the Taliban had forced into a remote corner of the country but never managed to eliminate. On 9 September 2001 *al-Qaedah* assassins posing as journalists murdered the military leader of the Northern Alliance, Ahmed Shah Massoud.

From their landlocked haven *al-Qaedah* orchestrated the September 11 attacks on the United States. The US was resolved to make an immediate response to the attacks, but the problem was how to strike into this remotest of regions. Utilizing hastily-established staging bases in Pakistan and Uzbekistan, special operations units from the various services worked closely with the Northern Alliance.

The first conventional forces into the country were the Marines of two MEUs, transported by C-130s and heavy lift helicopters. Staging through the port of Pasni in Pakistan, the first task on 1 December was to secure and improve Objective RHINO, a dirt airstrip that would serve as a stepping stone to the primary objective—the big airfield at Kandahar.[8]

After days of routine patrolling, the first action for the LAVs included in the two MEUs began in the early evening hours of 6 December when observers noted Taliban fighters probing the perimeter at RHINO. The ensuing fight was indecisive, but resulted in increased patrolling by the LAVs.

The primary mission was to intercept Taliban fighters fleeing southward ahead of the rapidly advancing Northern Alliance forces. Efforts on 4 December by mechanized Task Force SLEDGEHAMMER (3/6 reinforced by seven LAVs, a CAAT team, and fourteen Humvees) to shift north cross-country to help heliborne forces interdict the main east-west Highway 1 west of Kandahar were slowed by deep sand and rough terrain that pushed fuel consumption through the roof. After fording the shallow Arghandab River the vehicles became entangled in a small town surrounded by paddies and irrigation ditches.

The only incident came when aerial observers detected three men armed with AK-47s riding in a pickup truck. They crept through the darkness into the village, dismounted, and walked around a corner only to come face-to-face with an LAV-25 on full alert. The men bolted on foot, and according to one officer, "I don't think they quit until they were four or five miles away." Local guides eventually described a path through the maze of the village, and the task force moved on to block Highway 1.

The task force arrived in darkness, but in time to intercept a column of Taliban fleeing westward out of Kandahar in an assortment of stolen vehicles. In a running night battle the Marines claimed 50 Taliban killed, but local villagers later claimed to have buried as many as 150 bodies.[9]

On 7 December the Taliban abruptly surrendered the city of Kandahar, and by 10 December Marines arrived by air to secure the embassy in Kabul, closed since January 1989. On 14 December a Marine task force including the ubiquitous LAVs seized a more practical prize, the big airport at Kandahar.

The cost of seizing Afghanistan had been curiously cheap for the Marines, marred primarily by accidents. The Marines—and the LAVs—settled into a routine of patrolling, made risky not only by the Taliban but by the ordnance of a past war scattered liberally about the countryside. On 16 December eight dismounted LAV crewmen were providing escort for a team of engineers clearing expended ordnance from a dirt road on the Kandahar airfield. One Marine stepped on an ancient mine. Three men were wounded—the first casualties for any Marines operating with the LAVs.[10]

The fighting in Afghanistan shifted east, into the mountains of Tora Bora, in early December. Marine ground forces were already being drawn down. But they would return.

NOTES

1 Mortenson and Relin, *Three Cups of Tea*, p. 257.

2 The history of Somaliland is long and complex. For a good summary, see http://en.wikipedia.org/wiki/History_of_Somalia

3 Estes, *Marines Under Armor*, p. 192–193; Mroczkowski, *Restoring Hope: In Somalia with the Unified Task Force 1992–1993*, p. 175.; Estes, interview with Major Mike Campbell, December 2013

The following section on the tank deployment to Mogadishu is taken primarily from these sources unless otherwise noted.

4 Turner, *Tanks with the MEU: A Team for Success*, p. 41.

5 For example, Turner, *Tanks with the MEU: A Team for Success*, p. 39.

6 The Great Game has been the subject of numerous books, most recently Hopkirk, *The Great Game: Struggle for Empire in Central Asia*. For a simplified summary, see http://en.wikipedia.org/wiki/The_Great_Game

7 For example, Mortenson and Relin, *Three Cups of Tea*, p.238–243.

8 Knowlton, *Troops Seize Airstrip Near Taliban Base* : Lowrey, *From The Sea—U.S. Marines In Afghanistan, 2001–2002*, p.59.

9 Lowrey, *From The Sea*, p. 157–165.

10 Ibid, p. 188–190.

CHAPTER FIVE

Into Iraq

*Five days or five months, but it certainly isn't
going to last longer.*—Donald Rumsfeld

*My belief is we will, in fact, be greeted as liberators. . . .
I think it will go relatively quickly, . . . weeks rather
than months.*—Dick Cheney

Your Majesty! The Hundred Years War has started!
—Purportedly a line from an old Hollywood film

T HE 11 SEPTEMBER 2001 attacks on the United States demanded retaliation, and it was clear that the parties responsible were operating freely in the failed state of Afghanistan. By 27 October Brigadier General James N. Mattis, the senior officer
at Marine Forces Central Command—Forward (MARCent-Forward) in
Bahrain, was already planning raids into Afghanistan. However force caps
imposed by inter-service rivalries limited the participation of the Marine
Corps to two MEUSOCs in the initial incursions into Afghanistan. The
brunt of the fighting would be borne by various special operations groups
and by allied forces of the indigenous Northern Alliance.[1]

By early 2002 the attention of the Bush administration had, for good
or ill, become fixed upon the Baghdad regime. The dissent among the general public as to whether Iraq should be attacked was mirrored—albeit in-
visibly to the public—by disagreement within the military. Secretary of
Defense Donald Rumsfeld was a great advocate of "transformation" and
the use of small, powerful forces. Wishful concepts that underpinned the

drumbeat for war were the assumptions that the invading Americans would be welcomed as liberators, and that an Iraqi opposition (presumably highly vocal and politically connected exiles waiting in the West) would immediately step in to govern the country. It was to be a quick in-and-out campaign. A more realistic assessment by military planners was that the regime would simply collapse over night, leaving a dangerous power void in which the small Coalition force would be inadequate to police the country. Iraq cold easily become another Afghanistan.

Even relatively simple planning was continually thwarted by the shifting political situation. It was uncertain until the last minute whether British forces would participate. Turkey at first refused passage to British forces, and eventually to US forces; the entire operation would have to be mounted through a narrow corridor from Kuwait. There were the usual concerns over air operations: the Air Force was confident that Iraq could be bombed into submission, while the Marines did not want to give control of tactical air support to the Air Force. The Air Force wanted a long preliminary air assault, while the Army and Marines wanted a short air campaign to retain some element of surprise. Given the need to maintain a Pacific-based force for contingencies there, the Marine Corps forces would be cobbled together from various formations.

Daniel Benz had been an enlisted man in the Army National Guard while a student, was commissioned in 1995, and served in a number of billets including tank company commander in LtCol Mike Oehl's 2nd Tank Batalion; in September 2002 he assumed command of the H&S Company. One of his first priorities was to establish a Security Platoon. "A lot of Marines get an attitude like, 'If our supply trains get attacked, our Marines will automatically know exactly what to do because every Marine a rifleman.' In my mind, that is bullshit. Rifleman does not equal infantryman." Sixty men were led by a spare lieutenant and an NCO with an infantry MOS. Unable to do convoy security exercises on a crowded base, the unit could only do "walk-throughs."

Some Marines were baffled by the AVLB bridge layers assigned to each company, but "What you find, though, is Iraq is totally crisscrossed with irrigation canals, so it was a good thing we had them."[2]

In December 2002, 1st Tank Battalion conducted an elaborate field exercise—STEEL KNIGHT 03—to test concepts for a tank-heavy task force.

Pre-deployment commitments of the 3rd AAV Battalion precluded their participation. "This dashed the battalion's hopes of working with Captain Matt Watt's Lima Company of the 3rd Battalion, 4th Marines in a mechanized environment. First Tanks, based on the enemy and potential missions, anticipated attaching Lima 3/4 to form 1st Tanks into a Tank Heavy Task Force for combat operations." The battalion was however able to integrate two Reserve platoons from 4th Tank that would backfill for platoons afloat with MEUs.[3]

Gunnery Sergeant Timmie L. Legett had just been reassigned from 1st Tank to the I&I (Inspector/Instructor) staff at 4th Tank, and found himself right back in his old unit. "Activating I&I staff members for combat, when possible, greases the wheels and makes possible a seamless integration of the reserves. Within a few days of arriving, my reservists were full-fledged, contributing tankers, mechanics, and communication Marines of 1st Tank Battalion." Some of the Reservists, though, wished for more formal training. Captain Michael L. Ferrell was a former infantry officer, now the Executive Officer of Delta Company, 4th Tank Battalion. He was assigned as Platoon Leader, 3rd Platoon, Charlie Company, 1st Tank Battalion. Ferrell had attended only the brief Tank Commander's course, rather than the more extensive Armor Officer's basic course.[4]

After Desert Storm Mike Mummey was sent to the Security Forces at Mare Island Navy Yard. "Just myself and two other Marines on the whole island. That was pretty nice." Eventually he became the S-4 (logistics) Chief for 1st Tank and participated in the BRIGHT STAR exercises in Egypt, where he learned how lavish Army logistics were in comparison to the Marines.

When the Marine tankers could not get fuel, he encountered a lieutenant he had trained at the joint Army-Marine Armor Officer Basic School at Fort Knox. "He said 'Remember me? I was in AOB such-and-such.'

"I said 'I don't remember who you are, but I remember teaching AOB.' He goes 'Well here, bring your trucks over here. We got fuel for you.'" Such relationships would pay handsome dividends in the near future. "Made me look like I knew what I was doing; I go back to the tanks with all this fuel." Mummey, now the battalion's most experienced tanker, coordinated exercises with small logistics support and medical units attached to the battalion.[5]

In late January the Marines commenced the daunting task of moving a force around the world, only to have their careful plans thrown a curve when Major General (commander of the 1st Marine Division) demanded to know from his staff ". . . why he sees more infantry arriving, when we need tankers over here."[6]

The available forces would determine the plan. Army formations, though they would constitute the major part of both combat and logistics components, were more suited to battle on open terrain. Army armored and mechanized infantry divisions had far fewer dismounts—infantrymen who could root the enemy out of urban terrain—than a Marine Corps division of comparable size. The Marines, with their amphibious vehicles, were also better suited to breaching water obstacles.

The ultimate plan was a complex hybrid. The air campaign would be brief. The Marine Expeditionary Force—a Marine Air-Ground Task Force built around 1st Marine Division—would secure the oil fields of southern Iraq and prevent another environmental catastrophe like the one in 1992. The MEF would advance up the densely populated Tigris-Euphrates Valley with its many water obstacles and towns. It was felt that the attention of the Iraqi military would quickly become fixated on the MEF. A smaller MEB from 2nd Marine Division, Task Force SOUTH (soon renamed Task Force TARAWA), would follow to secure lines of communication and neutralize any forces bypassed by the MEF. British forces—the 1st United Kingdom Division (Armoured) and Royal Marine commandos added at the last minute—and a Marine MEU would secure the al-Faw Peninsula and the port of Umm Qasr between Kuwait and Basra. The main effort would be by the Army's V Corps, moving across the open deserts of western Iraq. Hard to reach northern Iraq would be the domain of special operations units, with airmobile Marines on call to secure the oil fields around Kirkuk. Major formations would converge on Baghdad, but uncertainty remained about who would claim that prize.[7]

The entire plan revolved around the widely misunderstood concept of "shock and awe." The phrase, catchier than the preferred Marine Corps "combined arms effect," does not imply the indiscriminate use of overwhelming violence. In fact, the sort of overwhelming violence practiced in Desert Storm would be counterproductive to the nation-building that must inevitably follow, alienating the Iraqi people by inflicting massive loss of life, and crippling reconstruction efforts by destroying infrastructure. Combined arms effect was instead the judicious use of psychologically paralyzing force. The concept, a sort of "new blitzkrieg," was to utilize better communications and superior mobility, multiple lines of operations simultaneous with parallel attacks, to keep the enemy off balance, presenting him with a torrent of irresolvable dilemmas.[8]

For the Marines a first priority was to seize the big pumping station at al-Zubayr. A cadre of experts from the British Military Works Force (the British company BP had built the facility) would be attached to 7th Marines. It would be a sort of hostage rescue mission: one ill-placed round could do billions of dollars in economic damage, and produce another environmental disaster.

Until the final moments, uncertainty reigned. The Coalition possessed unparalleled intelligence as to Iraqi equipment and capabilities. What remained unknown were intentions. Would the Iraqis fight? Would they employ chemical weapons? Did they even have a coordinated defensive plan?[9] There was still concern over Coalition proficiency and to what extent recent experiences had actually been absorbed. Despite recent experience in Somalia and Kosovo, most Marine infantry had no practical experience or training in working with tanks.[10]

Over a period of weeks Marine Corps assets and personnel flowed in: amphibious task forces and chartered airliners from the United States, MPS vessels from Diego Garcia in the Indian Ocean, and ferried aircraft.

Experienced tankers suspected that spare parts would as usual prove to be a limiting factor. Captain Dave Banning was a graduate of ROTC, and commander of Alpha Company, 1st Tank. "Before we deployed, my maintenance chief, who was an old crusty master sergeant who'd never spent a day in the Marine Corps away from a tank, was going to squirrel-hole a lot of parts. He wasn't shy about it. He kept everything out in the open. He didn't try to hide anything like a lot of the guys would. He laid them right out there and if it contradicted a Marine Corps order, he would ask, 'Do you want the tanks to run or do you want the order to be followed?' Initially we were told not to take any parts because we'd get everything we needed in theater. Yeah, right. Then we got word in the last week before we left that we needed to bring everything we got. So we took our own M88 and a couple Humvees. He jammed every part he could get his hands on into those vehicles and had them organized. Everything was very meticulously labeled. When we were detached from our parent battalion, the only thing that saved us were the parts he had squirrel-holed away."[11]

Personnel would be drawn from both the active duty and Reserves, mixed within the same units as empty billets were filled at the last minute. In some cases personnel would join their units within days of the beginning of the campaign. A Reserve unit—Major Bill Peeples's Alpha Company, 8th Tank Battalion—was all too typical. Peeples was commissioned in 1991; originally in logistics and maintenance, he left active duty in 1994. He did not become a tanker until 1998, and a tank company commander in 2000.

Roger Huddleston enlisted in 1977 and was an artilleryman, then was a TOW gunner in an Oklahoma Reserve unit and water treatment operator in civilian life. All his TOW sections were activated separately, and he was transferred into Alpha, 8th Tank at the last minute. The company First Sergeant billet had been empty for eight months. Peeples had talked to Huddleston only once on the telephone, and did not meet him until the unit was activated.[12]

Ordered to report to their home facility at Fort Knox, Kentucky on 9 January, the tankers drove through the night to report to Camp Lejeune North Carolina at 0700 on 11 January. Many critical billets, from platoon leaders to maintenance personnel, remained unfilled.

At many levels the planning and preparations were far from smooth. Alpha Company and the TOW unit were to be integrated into Task Force TARAWA, the cobbled-together 2nd Marine Expeditionary Brigade.

On the long ocean voyage Alpha Company was split among four ships, USS *Ashland* (LSD-48), USS *Gunston Hall* (LSD-44), and USS *Portland* (LSD-37), and the USS *Ponce* (LPD-15).[13] In some ways the ships were as decrepit as the tanks; *Portland's* propulsion system failed in mid-Atlantic, delaying its arrival for days.[14]

The company had left its own meticulously maintained M1A1 tanks in Kentucky, and took over vehicles culled and left behind by the 2nd Tank Battalion at Camp Lejeune. Once aboard ship the company found that the tanks were worn and beaten up; according to the unit records, ten of the fourteen were deadlined (inoperable). Peeples and his XO, Captain Scott Dyer, felt that the regulars who were overseeing the preparations for war regarded the Reservists as whiners and complainers.[15]

Maintenance chief SSgt Charlie Cooke and his men found tanks missing key components, and cannibalized some to keep other vehicles in operation. One Reservist—LCpl Randy Whidden, a civilian electrician from Florida—was a last-minute addition to the unit who would prove invaluable. The External Auxiliary Power Units (EAPUs) in particular were in appalling shape.[16] Bulky generators mounted on the back of the turret behind the tank commander's position, the EAPU provides power to operate critical systems such as radios and thermal gun sights without either consuming excessive fuel or draining the tank's batteries.

Despite their own problems the Navy came through again, manufacturing critical parts in the ship's machine shops. "We spent a lot of the trip over with all the Marines in the well deck, trying to get an idea about all the tanks. We were a little nervous of course because we hadn't seen the tanks. None of my Marines had been on them, seen them, or done anything with them. We were able to fabricate a lot of parts on the ships themselves, and fix them up somewhat."

Huddleston was particularly impressed by the crew of the USS *Ashland*: "The commander there, I believe it was Commander Howard, the ship's Captain, they was all for the Marines. Anything the Chiefs could do for us in their shops, they did." Upon arrival in Kuwait the company had fourteen mobile M1A1 tanks. Without specialized parts the repairs were limited, and many tanks were still not fully functional when the ship docked in Kuwait.

The Kuwait docks were so crowded that everything went ashore by LCAC, but once ashore they were able to scrounge critical parts from other Army and Marine battalions, and Huddleston ". . . could hear them (M) eighty-eights going all night long. They're working through the night and then during the day."[17]

Early January of 2003 found Mike Mummey back in Kuwait, preparing vehicles from the MPS ships and the most critical Fly-In Equipment (FIE). Inexplicably, all the essential refueling trucks were missing; they eventually arrived after gear shipped much later.[18]

Regular units like 1st Tank were better prepared than 8th Tank's organizational orphans, but Mummey admitted that "The only parts [things] that saved us, Ed, was our mechanics—before we left—they did the right thing. They were keeping unofficial parts on hand, and they embarked them over there. So we were living off the stuff we weren't supposed to have. And off the stuff we could scrounge from the Army."

Sam Crabtree, now a Reserve Staff Sergeant and senior turret mechanic, had gone on active duty as an I&I staffer, and was activated to backfill one of H&S Company, 1st Tank's empty billets. He arrived to find tanks already ashore in a staging area. This time around, "We knew a couple of months in advance we were going to get activated and go to 1st Tanks."

The Army's well-stocked depots in Kuwait were a godsend to the Marines. Mummey went to the Army facility at Camp Doha: "They got civilian contractors and Pakistani laborers in there working on these tanks for that equipment pool that the Army had.

"I go in there and I say 'Hey, my name's Mike and I'm in the Marines, and blah, blah,blah. I need a two breech operating handles, I need a firing probe—fire control stuff. They're like 'No problem. Hey Habi, go open up the locker for this man. Give him whatever he wants.'

"The Pakistani laborer, he opens up the wall locker and I just start grabbing stuff I need!"

There was no accounting. "He says 'No, man. We got stuff coming out our ears down here. Don't worry about it.'"

This was in sharp contrast to the Marines, since "We couldn't even get engineer stakes at first to set up [barbed] wire. We had to go down there to Camp Doha and get it from the Army."

Even so, some other units were envious. Daniel Benz recalled that "1st Tank Battalion, they started out with 66 tanks. Normally the number is 58. Somehow their S4 was slick enough to get them an extra eight tanks. Good for them, because things break."[19]

Like Task Force TARAWA, 1st Marine Division's heavy armored punch would be organized as two mechanized task forces. Task Force TANK (call sign TIGER) was built around Charlie and Delta tank companies, L/3/4 infantry, with Scout and TOW platoons in Humvees, and support units. A similar infantry-heavy task force was built around Bravo Company tanks, and the bulk of 3/4's infantry.[20]

Captain Seth Folsom, the CO of Delta Company, 1st LAR had just returned from a lengthy Pacific deployment with an MEU, and believed that the fragmentation of the battalion, with at least one company always away, harmed unit cohesion. In fact, the battalion would go to war without Alpha Company, already committed to an MEU deployment. Further, three companies (H&S, Weapons, and Charlie) would go by slow sea transport across the Pacific and Indian Oceans, while Delta Company would fly to Kuwait and utilize vehicles stored aboard MPS ships. Third LAR was similarly fragmented; B Company was on Okinawa, so B/4th LAR would be attached and re-designated E Company.

The grueling schedule was brutal on families. Folsom would be home less than three weeks before leaving again for war.[21]

Both active duty and Reserve Marines had to suffer through a battering whirlwind of last-minute minutiae: immunizations, replacing missing or damaged 782 gear, naming insurance beneficiaries, drawing up wills, and the inevitable last-minute marriages.

Delays, confusion, and shortages still reigned after arrival in Kuwait. For Mike Mummey, "The only thing we didn't have—and it's a damn shame—was the parts. . . . We didn't have spare parts for anything.

"A lot of stuff that we should have learned in Desert Storm, we didn't learn for Iraqi Freedom. Luckily, because of being at schools with Army guys, there were soldiers that I knew. . . . I could go find them, and they were like 'Here, Top, take this, take that.' You'd fill up the truck with hydraulic fluid, or turbo shafts, or get an O-ring set for the gunner's fire control handles—whatever stuff you needed. And the mechanics were doing the same thing. We would go down to Kuwait City to the Army base there and just go in there and beg and scrounge all we could."

Even basic functions like boresighting the main guns were limited by ammunition shortages—two rounds each of sabot and the new MPAT.

Yet despite all the problems, Mummey and many others felt that the situation was vastly improved over the Desert Shield buildup, since "There was more experienced people. Just like Diggs said, 'There wasn't nobody had done the do' before; knew what to expect. This time we all knew.'"

Peeples's company from 8th Tank had at last put all fourteen tanks into running condition, though five still had non-functional fire-control systems, and correct boresight could only be confirmed for ten.[22] Like most units Alpha/8th Tank had to wait until heavy transport trucks became available for the long trip to the forward staging area. The Reservists whiled away the time begging, borrowing, and sometimes literally stealing more parts in a final attempt to bring the tanks to full operational condition. New personnel also continued to be integrated. Major Scot Hawkins was added as a Forward Air Controller (FAC) and Captain Romeo Cubas was brought in as Platoon Leader for 3rd Platoon. This arrangement gave each company of the task force an artillery Forward Observer (FO), but B/1/2 was the only element without a FAC, an arrangement that would inadvertently lead to disaster.

Peeples's signal success was in arguing to organize his tanks as a single unit with attached infantry within 1/2, rather than diluting the unit's power by attaching a platoon of tanks to each infantry battalion. The tanks were organized into Team Tank (seven tanks plus a rifle platoon from B/1/2) and B/1/2 was designated Team Mech (four tanks, plus the balance of B/1/2). All 1/2 infantry was mounted in AAVs that would have to serve as personnel carriers; some of the regiment's remaining infantry would ride in trucks, sandbagged to provide minimal protection against mines and RPGs.[23]

Units like Folsom's LAR company might have made the trip in style aboard chartered airliners, but they faced problems that would have seemed familiar to armored crewmen of any previous war. Units were reorganized, vehicles stored aboard ships had to be extensively refurbished, there were no spare parts, new and complex gear like the Marine Data Automatic Communications Terminal (MDACT, a computerized battlefield mapping and display system) had to be installed and tested. A final indignity was that after 18 February the company had to share its LAVs with the newly arrived A/4th LAR, a Reserve company with no vehicles.[24]

With engineers struggling to supply the basics, the troops were left to shift for themselves in many ways, and mail-order was a new boon. Daniel Benz: ". . . you get one of those solar showers, and enough people brought those and water was never a problem. All the bottled water you could drink. They also had water buffaloes that were totally potable, so we never had a shortage of showers. We used some crates and whatnot to make our own."[25]

Bill Hayes enlisted in July 2001. All his recruiter's infantry slots were full, so the recruiter unilaterally switched him over to tanks. "I always had a bit of wanderlust. I'm from Georgia and everyone I knew was going to The University of Georgia; it seemed boring. I didn't really know what I wanted to go to school for. . . . I thought it might be kind of fun to travel and do exciting stuff. I got to do the exciting stuff. Didn't really get to travel. Went to Iraq a few times. It was a lot more exciting than going to school." He joined 2nd Tank in March 2002. "Even then there was a lot of cynicism about whether we were ever going to do anything."

Junior officers were in no better position. Dan Hughes had intended to join the Air Force straight out of high school, but the recruiter did not work on weekends. The Marine recruiter struck up a conversation, and ". . . next thing you know the Marine Corps sounds a whole lot better." In 1995 his CO suggested he sign up for a program that offered enlisted Marines the opportunity to attend college and be commissioned as officers. At thirty years old, and a former LAV crewman ". . . walking didn't seem like a whole lot of fun anymore. . . ." so he went into tanks. Upon arrival in 2nd Tank he found preparations underway for deployment to Iraq, but was fortunate to have a high level of experience among his men.

As loader in Sgt Erickson's tank, LCpl Bill Hayes was completely in the dark. "The leadership was like 'Oh, we might be doing something, guys! I can't say what I know, but we might be doing something. That's all we heard for like a month. . . . Then finally they're acting all weird, then 'Okay, guys, we got orders to deploy.' We kinda grabbed all our gear stuff and got on a plane and flew over there to Kuwait."

Arriving by aircraft before their tanks, "We just slept out in the Kuwaiti desert for a few weeks. Nothing to do, no vehicles. Very primitive conditions." Then "It's back to the guessing game again."

When the tanks did arrive from MPS ships, Dan Hughes noted that "some of them worked pretty good, some of them couldn't even fire up."

Gear continued to arrive until the last minute. Peeples: "Stuff seems to appear out of nowhere. . . . All of a sudden stuff like mine plows, and air panels and all those other things just seem to appear out of nowhere. Nothing that we really asked for, stuff that has just come."

Some gear like Blue Force Tracker was not scheduled to be fielded for years, and there were no installation instructions."In a manual there was a picture of a Blue Force Tracker in an M1A1, but the gear that we had gotten was not installed in the tank; it was installed in the bustle rack. . . . It will be useable to track my XO's tank around the battlefield. . . ." but the equipment would be of no use to the tankers sitting in a bag in the bustle rack. Other gear was cobbled into the tanks through the efforts of communication specialists.

While the press and diplomats fulminated, the military planners knew that war was inevitable. As in August 1914, once commenced it was virtually impossible to stuff the *djin* back into the bottle. From 15 until 19 March units began to move into forward positions, and engineers were busy creating preliminary breaches in the Iraqi border defenses—an anti-tank ditch, a ten-foot high earthen berm, and an electrified fence. A-Day air attacks would commence at 2100hours on 19 March, with G-Day ground attacks on 21 March.[26]

For Marines on the ground, the first indication that plans were again going awry came with the impact of a SCUD missile near Camp Commando in Kuwait at 1045hours on 20 March.[27]

A series of chemical attack alerts followed. During one such alert Folsom made an error of judgment: he forgot to spit out his chaw of tobacco, and faced an agonizing decision. Unmask and risk chemical death? Swallow the tobacco juice and face certain sickness? He finally decided to spit into the mask, nearly drowning in the process. With no way to verify that no chemical agents were present, the Marines fell back on a method dating back to World War I. A few junior men were ordered to unmask; if they survived, the all clear was given.[28] The frequent alerts exhausted the Marines as they donned and doffed the bulky suits and stifling masks. Many ceased to take the warnings seriously.

Again, the plan changed. Even before the assault began, the infamous "fog of war" set in. At 1845 RIPPER CP informed 1st Tank that the Republican Guards' *Medina Division* with 180 T-72 tanks might have moved into position along Highway 8. Battalion planners worked frantically through the night to realign forces. 1/7 and 3/7 would hold at the Line of Departure, while Alpha Tanks (reassigned from 3/7 to 3/4) would breach the border, then turn right to take any such enemy concentration in the flank.[29] Could the Marines commence an attack on four hours notice? At 2127hours the MEF transmitted the order to invade Iraq.

Bill Hayes was on night guard duty when the battalion Master Gunner came looking for the CO. "I knew right there something's going down. . . . It was literally within an hour they go 'Alright, wake everybody up. Let's go!" We got all the tanks ready to go and got'em in a big line."

Attached to BLADE (3/7), Captain Dave Banning, the Alpha Company 1st Tank CO, ". . . had just left the battalion command post where I had received the final op order for moving. I got back to my company assembly area and received word to send my tanks up to the berm to provide security. I hopped on the tank and I saw my TOW platoon take a pretty good incoming barrage. I pulled them back behind the Kuwaiti side of the berm and we drove our tanks up there as quickly as we could." With no advance warning, the company was to provide security for breaching operations by Kuwaiti engineers.[30]

Banning was told that ". . . they would like a security blanket to cover them. We parked a platoon of tanks up on the observation berms. My company XO and I went up there as well right as the sun was going down on the 20th. There was nothing we could really do for them other than moral support. . . ."

In the rush there had been no time to put thermal viewers on, so Banning took the opportunity to pass his viewer down to the driver, an operation that necessitated turning the turret to the rear, and elevating the main gun so that it was not pointed at friendly troops.

"I got on the intercom to find out what was taking my driver so long, I was just getting back up in the turret and the tank rocks. I immediately looked at my gunner and asked him what the hell he was shooting at. It felt like the gun had recoiled. I looked down and he said it wasn't him. I looked around the turret and everything looked normal except that my loader was down there with his head in his hands. I looked out on the back deck and I saw that it was on fire. The one-in-a-million Iraqi mortar shot got lucky, I initially thought, and put one in our back deck. Then I saw the amount of fire, realized that it wasn't a mortar round and told the crew to evacuate the tank."

Sprinting for the relative safety of some engineer vehicles 150 meters to the front, Banning could not find his driver. He returned to the burning tank, but the driver's "hole" was empty. Running back to the Humvee, he located the driver and found a hectic scene: the engineer officer was screaming into the radio. Banning took the radio, "turned it to my company tactical command post (TAC) and told everyone that I was all right. Everyone was fine and no one was seriously injured. I told them I wanted the platoon sergeant's tank of 3rd Platoon to be ready to execute the bump plan. In the meantime, I was talking to battalion and there was some confusion about whether the air wing knew we were down there and whether they thought we might be in an active kill box. We pulled all the tanks back behind the Kuwaiti berm and I told my maintenance chief to get ready to go recover the tank." He found that his loader, also the artillery forward observer, had been slightly wounded in one eye by the blast, but was fit for duty.

At the battalion CP "We found out that what hit my tank was a Hellfire from a Cobra. We could hear the helicopters when they arrived on station and my FAC didn't have the tactical air direction (TAD) nets. From what I understand, he had been told by battalion that since he wasn't controlling the air he didn't need the TAD nets, basically stiff-armed him."

Banning made sure the kill box was turned off. "We were told we had about an hour window to try and go recover the tank. So we went back up and got with my maintenance chief. We took 3rd Tank Platoon, my recovery vehicle and the Javelin platoon and we went back up, established a hasty perimeter around the tank and pushed the Javelins out far enough so we'd have some early warning if anything was going to happen. We got the M88 up there, hooked up the tank and ended up pulling it in a circle. One of the tracks was completely locked and wouldn't turn. At this point, we were getting close to our hour time limit that we had before we had to get out of there. The sound of gunfire was increasing but we still weren't taking any effective fire. We hopped on the tank, it wasn't burning anymore, took off our helmets, gas masks and our weapons. I got my artillery observer's radio and binoculars and we left it. We went back and tasked my TOW platoon to maintain visibility on the tank. I didn't want them to physically occupy it but I also didn't want the Iraqis to be able to get down there, start celebrating on it and turn it into some sort of propaganda type thing."

To add to the confusion, ". . . we were told that the scheme of maneuver had changed. Intel had reported that the *Medina Division* under the cover of a big sandstorm had been able to move their tanks down around Safwan. Now we were facing a division-size element of T-72s. I remember thinking, 'Holy shit. Maybe these guys did learn something over the last 10 years.'" Now ". . . everything we had rehearsed—going through Safwan and reducing the obstacles at the border crossing—was all out the window" and the task force had to be reorganized on the fly. "I left my tank retriever and two of my tanks with mine plows attached to 3/7. Everybody else ended up being attached back to 1st Tanks. This was a huge emotional event. I had to pull my tank platoon back from the infantry company that he was cross-attached to, give my infantry platoon back, my TOWs back and my Javelins back. It was a huge hassle." The injured FO went to the XO's tank, and Banning grabbed a spare tank crewman as loader.

While Banning made another trip to battalion, refueling went ahead of schedule, "So when I came back out my company was gone. I was sitting there in the desert with sand blowing all around me wondering where the hell my company was. I looked around, got on the radio and got linked back up with them. I'm writing out the operations order and at this point it was about 0200. We were supposed to move out at first light. I remember sitting there in my tank commander's hatch scribbling and the next thing I remember is my XO knocking on the hatch asking me if I was okay. He was a lifesaver. We got everybody spun up, got the company started and we were just following along behind Delta Company like ducks in a row."

In Task Force TANK's sector the Scout and TOW Platoons moved up to protect the breaching operation, but found to their surprise that that breach was to be conducted by local civilian bulldozers. Sporadic Iraqi fire caused a temporary suspension, but never deterred a crowd of civilians who gathered to gawk. After repeated warnings by the Civil Affairs team, they were finally dispersed by non-lethal fire.[31]

At other locations engineers began their own complex breaching drill, but found the mines and defenses far less formidable than those encountered in Desert Storm. Reconnaissance units mounted in Fast Attack Vehicles moved ahead to scout paths. The plan had changed at the last moment, and the main force moved through ahead of the LAR screening force. Trying to stay out of the path of RCT-5 and the attached 2nd Tank Battalion, Seth Folsom marveled at the sight as "score upon score" of infra-red beacons as the dark shapes of M1A1 tanks and AAVs thundered through the positions they had occupied only minutes before. First LAR's movement was repeatedly delayed, and the men slept under a constant roar of outgoing artillery and rocket fire.[32]

At 2042 the first Marine tank crossed the berm into Iraq.[33]

Bill Hayes's B Co/2nd Tank was attached to 1/5, which exchanged an infantry company to form two task forces. When they moved out, "It was really dark, and everyone had their IR lights on everything."

Arriving at the breach, "There was just like a breach line right out of a CAX exercise. They had gone in there with bulldozers, and they had marked it with chemical sticks on each side so you knew where the right side and the left side of the road were. . . . You just drove across and you could see the international border . . . where the Iraqis had their huge tank ditch. It was *huge*, and it didn't really do much to stop anyone."

Hughes' C Company, 2nd Tank was assigned a blocking position west of the Rumallah oil field, so the initial stages of the entry into Iraq were anticlimactic. The company had one tank damaged by a leftover mine from the earlier war.

At 0900 on 21 March 1st LAR at last moved through the breach and into Iraq. Among the units was a company-sized unit of LAV-AD air defense vehicles under Major William W. Johnson. With little air threat anticipated, the company carried no Stinger missiles, only its 25mm Gatling cannons. The three platoons, parceled out among the 1st Marine Division's RCTs, were destined to see little action, and the LAV-AD was later retired from the Marine inventory.[34]

The report of the massive force of T-72s proved to be a rumor, and there was only sporadic contact. Banning: "Delta Company ended up shooting some old rusted T-55 somewhere as we were going through. Charlie Company engaged some dug-in tanks along Route Tampa as Delta Company came on line to their south. When we came on line to Delta's right (southern) flank along Route Tampa, it was a nice, clear sunny day and we didn't see the *Medina Division* anywhere. Regiment decided that everyone would reattach in their previously designated task organization and continue the attack as previously planned. So now, mid-stride, we are detaching from our battalion, reattaching to another battalion and reintegrating all of our attachments that we had previously while we were on the move and potentially in contact. It was pretty chaotic."[35]

Seizure of the Az-Zubayr pumping station, cause of so much anguish, was relatively uneventful. Banning's tanks ". . . ended up passing south of the pumping station and meandering through all these fields -- tanks and amtracs crashing over these gigantic berms. It was a real rats in a maze type thing. I was tracking the company's progress between the military precision lightweight GPS receiver (PLGR) and my photo map of the Az Zubayr. Our route was traced out in pretty explicit detail where we were going. As we were going up the road towards the pumping station, my lead platoon commander's tank all of a sudden, the turret swings over, the gun max depresses and fires, and destroys an Iraqi tank about 20 meters off the road. The thing was dug in. You couldn't even see it. The entire tank was below ground level. You would have had to step down to step on the top of its turret. The tank right behind him traverses over and whacks one on the other side of the road. I look to my left and there was another tank about 20 meters away. I didn't shoot it because, at this point, the geometry would have had me shooting straight across our boundary into the flank of 1st Tanks. Given the friendly fire incident, I was very sensitive to angles of fire, geometries of fire and the effects of weapons systems past the boundaries. My XO, though, right behind me ends up popping it. So there's this tank cooking off right next to me and two tanks cooking off in the front. The battalion commander was on the radio screaming at us to hurry up. My concern at this point was that we could drive past these tanks, leave them and it wouldn't bother us at all, but these T-55s would tear the amtracs up. So I just didn't listen to the radio, told my unit to do a detailed search and to make sure there was nothing else out there that was going to whack them. It took about 10 minutes but it felt like an eternity."[36]

In the capture of um-Qasr the tanks and infantry quickly developed a drill for attacking sturdy masonry buildings using the new MPAT rounds. Major Ben T. Watson:

"A few times we just had the tanks run over a few machine-gun nests and just cross-steer crushing the guys beneath them. The new MPAT round, replaced the HEAT, is great for urban combat. We normally have the tanks create breaches for our asssault companies to enter buildings. 1 or 2 MPATs will create a hole big enough that you can drive an IFAV through. We have also conducted the now patented Tank / MSPF hard hit assault. I am sure the guys at SOTG would be crying because we broke all the rules. We had to take down the Ba'ath party HQ in Umm Qasr. We did it with the tank platoon, force recon, and the trailer platoon. We lead with tanks, the four tanks got on line and blew the crap out of the building with their main gun using MPAT and created two breaches. Once the trailers dismounted and moved abreast of the tanks they switched to 7.62 and .50cal hosing down the house. When the trailers were ready to move forward we shut the tanks off and the trailers secured the perimeter of the house. Tanks were then again pushed forward. A section covering each incoming road. The force platoon went inside and finished the clearing operation. The biggest takeaway was the tanks work great in MOUT."[37]

At mid-day A/8th Tank took its turn to move through their assigned gap—without one tank (SSgt Jeffrey Filipowski's)—that had broken down. The inevitable delays in pushing an entire division through the narrow breaches led to successive delays as units were told to hold in place.

Within eleven hours the lead element, the 5th Marines RCT, had seized most of its objectives. The Iraqi *51st Mechanized Division* had simply evaporated, and even sabotage was minor. The first forces cleared the road north, and C/1/7 passed through and successfully seized the al-Jubayl pumping station.

The overall plan called for some units to screen or hold in place. Third LAR would feint, then withdraw and Team MECH with Banning's Alpha, 1st Tank was to occupy a large military compound and secure the road toward Basra. After seizing the unoccupied compound, Banning was concerned about an open flank and unsecured areas to his rear. But the enemy was even more confused.

"About the time we start taking fire from the rear, the dismounted infantry platoon started returning fire and was trying to figure out where in the hell it was coming from. I looked at the map again because I was very concerned about the direction of fire. I didn't want to shoot into friendly forces. My loader, Lance Corporal Chambers, elbows me and says, 'Sir, I didn't know the Army was up here.' There were three or four trucks coming up the road and two of them are towing 57 millimeter anti-aircraft cannons. As I looked at these trucks, they looked the same as those flat-nosed Army trucks but in a different color. I was sitting there thinking that the Army wasn't supposed to be here and did I miss something in the plan. I did the mental run through to make sure they weren't Army trucks. They weren't and so I gave the fire command to my company; and as soon as I said fire, my first platoon commander had a sabot round off like a shot. You could see it streak downrange and it hit the lead truck right in the cab. You see the spark and a little bit of smoke. All the other trucks behind him were going crazy. All my other tanks opened up now, and in about 10 seconds all the trucks are smoking ruins."

Iraqi soldiers spilled out of the trucks to shelter in bunkers, some occasionally emerging to try and bring their cannon into action, ". . . and they would get whacked too. About this time, the battalion commander showed up. My first sergeant was putting the collar on the infantry platoon and telling them to back off." Without adequate infantry, and faced with an aggressive infantry officer, Banning's mind kept going back to ". . . that book, *We Were Soldiers Once and Young,* where that lieutenant charges off and gets his platoon cut off. That was one of my overriding concerns. I didn't want to drive up there and get embroiled in something that the battalion was going to have to figure out how to get me out of." After a mechanized infantry company arrived, the Marines seized the building complex without incident, and then sat through an on-again-off-again series of alerts about enemy counterattacks. "We waited for the Brits, the 7th Armored Division. They were coming in to do a relief in place (RIP). As soon as the danger of any kind of attack across the canal vanished, we did the changeover, we pulled back, refueled again and then got ready for the road march up to the next assembly area. That was our first day.[38]

Thirty kilometers to the rear, Task Force 3/4, with Captain Brian R. Lewis's Bravo Company tanks, was struggling to catch up with the main tank task force. The tanks were in the lead. Lewis:

> We started the movement in a staggered column and found the terrain to be restricted enough that we were basically restricted to movement on the hard surface road. Red platoon, Lt Chad Hall's platoon was in the lead, with HQ, White and our infantry platoon, with our M-88A2, in trace. Red reported that "things are getting weird up here." Before he finished that call over the company net all hell broke loose, and main gun, .50 Cal, coax and loaders 240 all started within a few seconds of each other. I think we were basically in a state of disbelief as to what we saw. One second we had negative enemy contact and the next there were enemy forces running around all over the place.

The task force had crashed head-on into a force of T-55 tanks, BMP-1s, French-made Panhard armored cars, and a reinforced infantry company, still full of fight.

LtCol [Bryan] McCoy (the infantry task force commander) was behind my tank and my tank leader was winging for him. I remember telling my gunner [Sgt Hahn] to engage the dismounts he was observing to our Northern flank, I popped my head out to look south and saw the Colonel's vehicle stop the driver and Colonel jumped out and started engaging dismounts 15 to 20 meters behind my turret. That's when I saw a T-55 not 15 meters to my flank. It was completely dug in with burlap camouflage and it was dug in to the point of not being able to engage with its main gun. The Colonel fragged (hand grenaded) the tank after fighting with the hatch, which was apparently combat-locked. My main concern was not for the Marines engaging enemy forces; it was a family of at least 15 civilians caught in between my tanks and the enemy. My tanks were firing all of their weapon systems at separate targets all along the company's column. After several minutes of firing, and more than one radio call to warn tank commanders of the location of the civilians now huddled together just off the road, we ran into Lima Co 3/4 attached to 1st Tank Battalion. Our fires and location prevented us from dismounting our infantry to clean up what was left of the enemy, so we turned northeast along Route Red and prepared for our next contact.

Lewis later mused that the situation was hard to accept "for about 1/10th of a second," but that ". . . the Marines instincts kicked in and they started killing the enemy aggressively and without prompting. Seeing their Battalion Commander dismounted, killing the enemy with them, was the foundation that carried Bravo Company through the war. Aggressive bare-knuckle, brute force and ignorance carried the day and the war for us."[39]

The hasty attack necessitated leaving some vehicles behind to relieve the glut along the routes of advance. Second Tank carried large rubber bladders filled with fuel slung on the gypsy rack. When the internal fuel cells were near empty the bladders could be dropped to the ground and the tank driven over them to squeeze the fuel into the internal fuel cells. If a critical need arose, additional fuel bladders could be slung beneath helicopters and brought to the tanks. Relying upon the Air Wing for support, the tank units left their fuel trucks, ambulances, and most other soft-skinned vehicles behind. Breaking loose from support trains, 2nd Tank outran even communications; the CO, LtCol Mike Oehl relied upon e-mails sent by satellite phone.[40]

On the first night in Iraq the Marines briefly settled into night defensive positions. Everyone's nerves were still on edge. Bill Hayes: "There was like some huge camel herd out there in the middle of nowhere. One of our tanks, there was some small explosion or something had hit it. No one could ever figure out what that was. We ended up spending thirty minutes looking out through (sights) . . . All you could see was this huge camel herd out there."

The original TARAWA operational plan had called for an assault to capture an-Nasiriyah, but the requirement had been downgraded to a "be-prepared-for" mission.[41]

Instead TARAWA was to temporarily stand aside for the Army. Peeples: "They're going to have to send their logistics train through Task Force TARAWA area so we have to wait first off for that logistics train to clear our zone." While establishing its defensive positions, TARAWA suffered its first loss—self inflicted when a tank, moving without a ground guide, ran over two sleeping Marines.[42] Captain Cubas also lost another of his tanks to mechanical failure.

In the 1st Marine Division sector the Lima Company infantry of Task Force TANK, supported by a platoon of tanks, fought for control of key bridges over the Sha'at al Basra water obstacle (Sha'at means river mouth in Arabic).

In the morning light of 22 March 1st Tank's lead elements found themselves intermixed with the enemy. With TOW gunners engaging tanks on the north shore of the river, a stay-behind group scored an RPG hit on an M88 of the Combat Trains. Captain Ruben Martinez:

> As I completed shaving, I saw an RPG smoke trail out of the corner of my eye originating from about 175 meters to the south of the M88. As I turned to look at the M88, I saw Sergeant Percy standing in the vehicle commander's cupola. The explosion rocked the M88, causing a fireball to come out of the vehicle and smoke to begin billowing from the vehicle. Master Sergeant Mummey, walking toward the vehicle, also saw the RPG fired and hit the vehicle. Corporal

Rugg, a crewmember of the vehicle, began to exit the M88, standing just outside the hatch and turned toward Sgt Percy to ask what happened. Sgt Percy shouted that he couldn't see. I could see Cpl Rugg's lower legs badly injured with many bleeding lacerations. Corporal Rugg then jumped down from the vehicle, obviously not realizing the extent of his wounds. He landed on his feet and immediately fell over yelling "Oh my legs!" As I ran toward him, I saw Sergeant Jones running toward Cpl Rugg as well and I turned to retrieve a stretcher from BAS 2. Chief Petty Officer Lugo directed BAS 1 to park near the M88. Upon returning, Sergeant Villela, Corporal Rodriguez, and Lance Corporal Burgos lifted Corporal Rugg onto the stretcher and I went over to assist Sergeant Percy off of the M88 and walked him over to my vehicle. I could hear sporadic small arms fire from our Marines and fired at our position. After washing Sergeant Percy's eyes out with water, I called in the contact report and the number of injured Marines. As I was on the radio, I observed Sergeant Jones pulling Master Gunnery Sergeant Denogean's limp body out of the burning M88, assisted by Corporal Rodriguez. What we would later learn, that Chief Lugo was inside the M88 passing the Master Gunnery Sergeant out of the vehicle. . . . Sergeant Jones returned and entered the burning M88 to remove AT-4 missiles that might have gone off due to the heat.[43]

The bridges were still under intense fire, preventing recon teams and British SAS advance parties from moving into the city. The SAS tucked in with the tanks, relaying information provided by an informant with a cell phone inside the city of Basra. Nevertheless, civilians continued to mill about in the middle of the fighting. A platoon of tanks from Delta Company was providing security at one of the bridges when three vehicles approached from the north bank. 1st Lt. Vincent Hogan:

We first observed the three vehicles when they stopped about 1500m to our front; the lead vehicle was clearly a MG-mounted 'technical.' As we went to REDCON-1, we observed 10-12 armed personnel on the 'technical'; the gunner put on a gas mask. Suddenly they were hauling ass towards our position. Red-2 shouted, 'RPG!' seeing one passenger lift an RPG out of the vehicle bed. We engaged with coax and .50 cal. In a minute or so, two vehicles were burning, with one remaining on the road. No one was moving. A couple minutes later, people started picking up weapons scattered from the 'technical'; we continued to engage. The whole engagement lasted maybe 10 minutes. After the fight, some of the gunners were able to observe through their powered sights 'TV' sloppily taped on the sides of the SUVs. I decided to inform the S-2 about this potentially useful, enemy vehicle marking. It didn't even occur to any of us that journalists would try to participate in an Iraqi suicide charge."[44]

The occupants were indeed three journalists and a translator employed by the British *Independent Television News*.[45] They paid the ultimate price for poor judgment.

The enemy continued to counterattack piecemeal. GySgt Steven Heath:

I saw two 'technical' vehicles come across the bridge with a T-55. On the near side of the bridge, about 2700m away, the vehicles had formed a wedge with the tank in the lead, two 'technicals' flanking. We saw them at an angle. The tank had guys riding all over it, World War II-style. Our first shot, sabot, went through the chest of a tank rider, and destroyed the 'technical' on the far side. The tank stopped and everyone jumped off. I fired the second shot: through a berm, in front of the tank, a hull shot. Then, a guy in black jumped in the driver's hole and started it. There was just enough time to see the white puff as the engine started and the tank began to move. Red-3 put a shot square in the turret, causing immediate 'secondaries.' Good section gunnery. It was almost like the Iraqis tried to adapt Soviet tactics to 'technicals,' like a Combat Recon Patrol. It wasn't working out too well for them.[46]

The SAS informant relayed information of an armored counterattack, which was broken up by helicopter gunships. The enemy had one more weapon, one that would have been familiar to Marines of six decades earlier—suicide bombers. Captain Greg Poland, the CO of Delta/1st Tank was providing security for the grunts of L/3/4 in the blocking positions outside Basra. The grunts pulled a man out of a car, and as Poland radioed a situation report to battalion ". . . the abandoned car next to my tank made a quick fizzling sound and then blew up. I only had enough time to buckle my knees and drop inside the tank. At this point, I wasn't bored anymore. Shortly following the car explosion, several bursts of small arms fire came across the top of my tank from the south, but it was sporadic and not well controlled. The best part of the night was cross talking with Capt Watt. At one point (after the car explosion), we were actually laughing with each other on the radio. As the enemy fire was going on around us, I said, 'Hey Matt, getting shot at really IS kind of fun, isn't it?' He offered his agreement and we continued through the night."[47]

By 1300 on 22 March the Marines were turning local responsibility over to the British, and TARAWA was headed west. The plan was to outmaneuver the Iraqi resistance with a fast advance and numerous feints, while the Army's V Corps raced across the open desert to the west and into Baghdad.

In the main body the advance was more like military tourism. Bill Hayes: "Being a loader in OIF-One was one of the best jobs in the military. You're the driver, you need to be focused the whole time. If you're the gunner, you're down there buried inside the tank. There wasn't really a lot of action in OIF-One. . . If you're the tank commander, you have the burden of responsibility. The loader, he just sits on the loader's hatch all day long, kind of minding my own business. Had a great view of everything. Once we stopped I had a lot of the work to do, but as long as the tank's moving . . . I can just sit there and enjoy the view."

There was a prescribed drill at every stop. "You always checked the hydraulic fluid, the oil, and if you have enough time you want to check the tracks, drive the tank backward slowly to make sure none of the bolts or anything have fallen off." The abrasive sand was brutal on the tank systems. "That track was just burning-up hot. . . . The track pads, you could tell were being pushed to the limit so far as heat."

After weeks the crews had settled in to live in the tanks. "It's like your home, your RV. You got all your water and you got all your junk food that's been mailed to you all stashed everywhere. It wasn't too bad. The people who had it bad—I would not have wanted to ride from Kuwait to Baghdad in the back of an amtrac, jammed in there with twenty people. You can't see anything. Very loud." When a tank crewman grew bored he could "See what was going on on the various radio channels. We kind of have a nice view of what's going on, where we're going."

The rapid pace of the advance also had its downside as numbing fatigue set in. Benz:

(On 24 March) We put up the radio. My first sergeant is getting a catnap. And the battalion was all set in a circle, all the tanks in a circle. We heard a *bop-bop-bop-bop!* First thing, we kind of hit the ground and, come to find out—and this is to give you sense of how tired your people are getting—Delta Company, our Reserve company, basically a guy was getting out of a tank and reaches out to pull himself up, puts his hand on the butterfly trigger, .50 cal. The damn thing was not on safe. Worse of all, [*inaudible*], over the side to check on the engine, next tank over, there's a Marine (LCpl Eric Orlowski) standing on top of the tank, *bop-bop-bop-bop!* . . . I mean, just killed him instantly. The bad thing was that we were the ones there first. Then the tanks rolled in. We had a corpsman but the surgeon, everybody else, is still out in the road, a major frickin' traffic jam. First thing, stay calm, urgent MEDEVAC. Only a medical officer can say this guy is dead. But he's dead. So the surgeon and the (S-)'4 are in the same Humvee, and they're freaking out on the radio because they're caught in traffic. So I call them back: "Hey, I don't want to get everyone else freaked out in the battalion. It's a 'routine' MEDEVAC." "What?!" "Listen, it's a 'routine' MEDEVAC." The thing was, it didn't really get in the battalion's loop. I was kind of shocked how the battalion commander—we had to be somewhere by 0600 the next morning. I got a 30-minute catnap.[48]

Back in the massive traffic jam to the rear there was little enemy action, so fatigue was the major problem. Despite his rank, Sam Crabtree was assigned a five-ton truck in the Field Trains, since there were few licensed drivers. "We were going so slow. . . . We would go five, ten miles an hour." In the massive convoy "A lot of vehicles were getting in accidents because people were falling asleep. . . . You get bored. You fall asleep."

On the large scale all was going well, with less than expected resistance. The entire American component had crossed on a fairly narrow front, restricted by the local road network, and would continue to follow the roads as far as possible to ease passage. But the plans were beginning to morph with the rapid advance. Near the Euphrates River V Corps would swing west, and TARAWA would seize the eastern crossings near the town, easing the way for passage of the 1st Marine Division. Unfortunately the massive logistical trains of the Army's 3rd Infantry Division of V Corps was caught in a huge traffic snarl south of the main crossings of the Euphrates near an-Nasiriyah, clogging Route 1 along the south bank. The intersection of Routes 1 and 7 south of the city was chaos as TARAWA, with the mechanized 1/2 in the lead, fought its way through the jam.

On a more personal level, Seth Folsom noticed that the Iraqi civilians merely gawked and waved at the passing parade. The Marines tossed out "humanitarian MREs," to the civilians, making note that adults would push children aside to grab the packages.[49]

No one had cause to suspect that, as often happens in war, erratic behavior of the enemy, and a navigational error by a junior Army officer would result in one of the most confused and savage battles of the war.

NOTES

1 Reynolds, *Basrah, Baghdad, and Beyond*, p. 4–11.

2 Galuzska, *Interview with Major Daniel Benz*, p. 5.

3 Anonymous, *Operation Iraqi Freedom Iraq—1st Tank Battalion*, p.3.

4 Ibid, p.3–4.

5 Ibid, p.4.

6 Ibid, p.6.

7 Reynolds, *Basrah, Baghdad, and Beyond*, p. 38–39, 66–67.

8 The concept has its modern genesis in Ullman and Wade, *Shock and Awe: Achieving Rapid Dominance*.

9 Reynolds, *Basrah, Baghdad, and Beyond*, p. 44–45.

10 Storer, *Tank-Infantry Team in the Urban Environment*, p. 61.

11 Lessard, *Interview with Major Dave Banning*, p. 15.

12 Bonadonna, *Interview with Major William Peeples and First Sergeant Roger Huddleston*. All direct Peeples and Huddleston quotes are from this interview.

13 *Ashland* and *Gunston Hall* were named after two of the original class of LSDs (LSD-1 USS *Ashland* and LSD-5 USS *Gunston Hall*) that saw service in World War II.

Both vessels bear names with historic ties to the Marine tank units.

14 Lowry, *Marines in the Garden of Eden*, p. 42. The unit record, Glover, H., *Command Chronology Report for Calendar Year 2003, Company A, 8th Tank Battalion,*, reports that the unit was divided among three ships.

15 Glover, *Command Chronology Report for Calendar Year 2003, Company A, 8th Tank Battalion*; Peeples painted a much less critical picture in Bonadonna, *Interview with Major William Peeples and First Sergeant Roger Huddleston.*

16 Lowry, *Marines in the Garden of Eden*, p. 35.

17 Glover, H., *Command Chronology Report for Calendar Year 2003, Company A, 8th Tank Battalion*; Bonadonna, *Interview with Major William Peeples and First Sergeant Roger Huddleston.*

18 *Operation Iraqi Freedom Iraq—1st Tank Battalion*, p.7.

19 Galuzska, *Interview with Major Daniel Benz* p.9.

20 *Operation Iraqi Freedom Iraq—1st Tank Battalion*, p. 8–9.

21 Folsom, *The Highway War*, p. 5–13.

22 Glover, H., *Command Chronology Report for Calendar Year 2003, Company A, 8th Tank Battalion*. Boresight confirms that the aiming point and the impact point actually coincide.

23 Pritchard, *Ambush Alley*, p. 21; Lowry, *Marines in the Garden of Eden*, p. 57.

24 Folsom, *The Highway War*, p. 23–30, 44.

25 Galuzska, *Interview with Major Daniel Benz* p.15.

26 Reynolds, *Basrah, Baghdad, and Beyond*, p. 62.

27 Because of fighting across time zones and nearly instant communications between headquarters in the US and fighting in the Middle East, most official records and some memoirs record events in Zulu Time (Greenwich Mean Time). Local time in Kuwait and Iraq was Zulu+300. Here all times are converted to local time for clarity.

28 Folsom, *The Highway War*, p. 38–42.

29 Anonymous, *Operation Iraqi Freedom Iraq—1st Tank Battalion*, p.10–11.

30 The following account of this action is excerpted from Lessard, *Interview with Major Dave Banning*, p. 5–7.

31 Anonymous, *Operation Iraqi Freedom Iraq—1st Tank Battalion*, p.10.

32 Folsom, *The Highway War*, p. 88–89.

33 Reynolds, *Basrah, Baghdad, and Beyond*, p. 68–69.

34 Ives, *Interview With Major William W. Johnson*, p. 4, 6.

35 Anonymous, *Operation Iraqi Freedom Iraq—1st Tank Battalion*, p.12; Lessard, *Interview with Major Dave Banning*, p. 7.

36 Lessard, *Interview with Major Dave Banning*, p. 7–8.

37 Watson, *Ramblings part II*, 14 April 2003

38 Lessard, *Interview with Major Dave Banning*, p. 8–9.

39 Anonymous, *Operation Iraqi Freedom Iraq—1st Tank Battalion*, p. 42–43.

40 Landers, *The Marines' 2nd Tank Battalion Used Speed and Armor to Make Quick Work of Saddam Hussein's Regime.*

41 Glover, H., *Command Chronology Report for Calendar Year 2003, Company A, 8th Tank Battalion*.

42 Lowry, *Marines in the Garden of Eden*, p. 109–110. The driver has such poor visibility that in darkness or congested conditions another Marine walks in front of the tank, giving the driver visual signals.

43 Anonymous, *Operation Iraqi Freedom Iraq—1st Tank Battalion*, p. 15.

44 Ibid, p.16.

45 Elliott Blair Smith, *Into Iraq*

46 Anonymous, *Operation Iraqi Freedom Iraq—1st Tank Battalion*, p.16–17.

47 Ibid, p.17–18.

48 Galuzska, *Interview with Major Daniel Benz* p.20.

49 Folsom, *The Highway War*, p.108. Humanitarian MREs include only "universal" foods that will not violate religious dietary proscriptions in most parts of the world.

CHAPTER SIX

Bridges in the Desert—An-Nasiriyah

Confusion in battle is what pain is in childbirth—the natural order of things.—General Maurice Tugwell, British Army

T WO CRITICAL PATHS of the invasion route brushed the grimy little industrial city of an-Nasiriyah before swinging west of Baghdad and into that city, respectively. The city consisted mostly of slums, petroleum industry facilities, and military compounds.

The Marines were wary of going into any city. Urban areas entangle and absorb military strength, pass the advantage to defenders who have intimate knowledge of the local terrain, and require slow and usually costly house-to-house or even room-to-room infantry fighting. In the Cold War era the US Army had made it policy to avoid urban combat. The Marine Corps—with Seoul (1950) and Hue (1968) still fresh in its institutional memory—had trained for urban combat, and was under no illusions. The subject of an-Nasiriyah had in fact been the subject of considerable controversy in Marine Corps planning. Logistics routes for both the Marines and V Corps passed through the chokepoint of an-Nasiriyah. The 3rd Infantry Division of V Corps would pass through the far western outskirts of the city along Route 1, but the final plan placed the city firmly in the Marine zone.[1]

The Marines would have to go into the teeth of whatever resistance the regime could muster. Route 1 was the best highway, but it was reserved for the use of V Corps. Route 7 branched off Route 8 and crossed the Euphrates River and the Saddam Canal in the city, linked with Route 16 north of town, and then veered north and west toward Baghdad. The bridge over

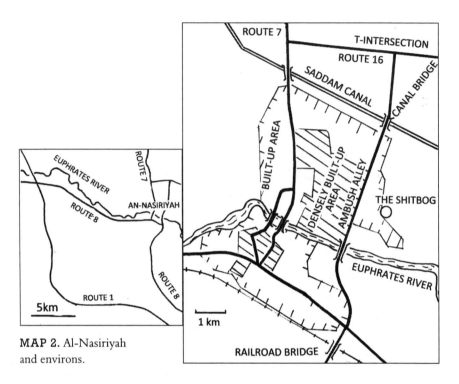

MAP 2. Al-Nasiriyah and environs.

which Route 1 crossed the river was about 10 kilometers west of the city, but it was under construction, narrow, and might not bear heavy traffic. The Marines would need to control the main bridges inside the city and on the direct highway.

With a mostly Shi'a population, an-Nasiriyah was at the best of times ignored by the Saddam regime. At other times the populace was viciously oppressed. In the aftermath of Desert Storm the CIA had encouraged a revolt. The Ba'athist regime had unleashed terrible reprisals, and it was still garrisoned by a force of regular Iraqi Army troops, *fedayeen*, *al-Quds* volunteers, and Special Republican Guards.

TARAWA's plan was carefully rehearsed. The main highway through the central part of the city—a 4500m long boulevard lined with low buildings—was already dubbed Ambush Alley. TARAWA would try to avoid becoming involved in street fighting by veering through the eastern margins of the city, but it would still have to capture the critical bridges. At that point RCT-1 would pass through TARAWA's lines to continue the advance on Baghdad.

Alpha Company infantry in AAVs would veer around the central city to seize the bridge over the Euphrates, Bravo would swing east around the fringes of the city to capture the Saddam Canal bridge, and Charlie would move north to capture a key highway juncture dubbed The T-Intersection. Tanks would provide mobile firepower to "shoot the infantry onto the objective." Securing the city—and the Main Supply Route through it—would have to wait until after the 1st Marine Division had passed.

On the morning of 23 March all was going according to this plan when at 0500 hours local time Charlie Company, 2nd LAR relieved Army combat troops west of the city, and TF TARAWA passed elements of the 1st Marine Division that were halted alongside the highway south of the city.[2]

Mike Mummey, with 1st Tanks' trains, was caught up in this snarl, with two M88s, fuel trucks, two thinly-armored Humvees ("We only had those because our embark officer was really shit-hot, and arranged for us to fly in a [new] Humvee from the States and trade it to the MPS ship for hard-back versions"), and heavy trucks with trailers laden with "whatever parts we had scrounged up. I had enough . . . to resupply one company if needed. We were always a few clicks behind the battalion as they moved."

Mindful of the confusion in the first war, he organized his own security team. "I had an MTVR, which is that seven-ton truck, and I had a couple of cooks, a couple of tank mechanics, the grunts gave me two grunts to be on it." Also included was a TOW section, "So I had four gun trucks—four TOW launchers—and one hard-back vehicle with a hard-charging sergeant. . . ." All told there were about thirty-five vehicles and 120 Marines.

Marine reconnaissance units crossed the Route 8 bridge in the western part of the city and penetrated into the main city, but withdrew after being ambushed. They reported to TARAWA that unlike most Iraqi forces, the ones in the city were in a mood to fight.

Like most, personnel of the Army's 507th Maintenance Company were sleep-deprived and exhausted by the heat. Several broken down trucks were under tow, and ten-ton wreckers towing broken-down HEMTTs with large trailers still affixed, made for unwieldy tows impossible to turn around and painfully slow to negotiate any sort of turn. The little convoy had fallen far behind the rest of their parent convoy, as vehicles from various units became intermingled in the vast northward flow. The 507th was supposed to pass to the west of the city, but had gotten off course taking a supposed shortcut. They could still have bypassed an-Nasiriyah if they turned left before reaching the river to follow Route 8 west of the city, but none of the officers or NCOs had a map. All were relying upon GPS "waypoints"—satellite coordinates—for the location of turns. A simple map could have prevented the ensuing disaster.

The Marines of TF TARAWA were drawn up alongside Route 8 south of a railway overpass, with an enormous logjam of vehicles behind. The column for the assault would have Team TANK and the Combined Anti-Armor Team, or CAAT[3] in the lead, then Bravo, Alpha and Charlie Companies, 1/2. Peeples's small command was already reduced by two tanks; one had tumbled into a hole in Kuwait, another was broken down and under tow far to the rear. The worn tanks were still not completely functional. The turret traversing system on SSgt Dominic Dillon's tank in 2nd Platoon was inoperable and the heavy turret had to be traversed by hand.

The tankers were not happy with the rubber fuel bladders that encumbered their gypsy racks. The bladders extended the range of the fuel-swilling tanks, but presented a fire hazard and interfered with turret traverse.[4]

The lead elements of TARAWA came under mortar and machine gun fire from buildings on either side of the road, and Cubas's tank became mired while attempting to move off the road. The tankers were engrossed in pulling it loose with another tank under fire when the sixteen-truck convoy from the 507th stopped to ask directions of 1st Lt. James Reid of Weapons Platoon, C/1/2. The Army convoy then inexplicably motored past the startled Marines and across the overpass toward the river. At least some of the soldiers in the convoy wondered why they were pulling ahead of the combat troops. They were obligingly waved through an Iraqi Army checkpoint.[5]

The defenders of the city were probably more startled than the American soldiers. Not a shot was fired at the errant convoy, though there was considerable dashing about by the defenders. It soon became obvious that the convoy was seriously off course. A truck ran out of fuel and was quickly refueled, but turning the cumbersome tows around in narrow side streets consumed precious minutes. Time quickly ran out for the 507th.

What ensued was more an urban gang fight than a battle. Iraqi irregulars dashed about wildly firing AK-47s and RPGs at the trucks. Some just stood at the roadside and wildly sprayed rounds at the passing trucks. Others blocked the convoy by throwing debris into the streets, or pushed vehicles into their path. Inaccurate as the fire was, there was lots of it, and American vehicles and soldiers were inevitably hit. The convoy quickly broke into three elements and the first two vehicles, one carrying the company commander, managed to speed out of the trap. Occupants of four more trucks were fighting for their lives behind. At the rear the largest group, including the slow heavy trucks, was in a hopeless position. One heavy truck ran off the narrow streets and was stuck with its rear in the road; a Humvee crashed into the stalled truck.[6] Weapons jammed from accumulated dust and fine sand, or ran out of ammunition. The Iraqis soon overran the survivors.

The Iraqis and foreign volunteers had recovered from their surprise, and at 0900 the advancing Marines of TF TARAWA were still under mortar, machinegun, and RPG fire from buildings in the muddy fields south of the railway overpass. Javelin and TOW gunners of the CAAT engaged Iraqi tanks, destroying nine.[7]

When Peeples's gunner tried to traverse the turret on "Wild Bill," the fuel bladder jammed the turret. Peeples leaned out and slashed the lines securing the bladder, dumping it into the road. The rest of the crews followed suit. Dyer saw an American truck that came south from the direction of the river, then doubled back. The Marines had been warned of the potential for enemy soldiers using American trucks and uniforms as ruses. The tankers were still unaware of the 507th's plight. The Humvee reappeared trailed by two more damaged trucks spewing smoke.

The 507th's commander, Captain Troy King, made it back across the river bridge in the Humvee. When Peeples dismounted and talked to the nearly incoherent King he gained only a rudimentary picture of what was going on north of the river, but immediately headed north. Dyer found the radio net jammed by superfluous chatter when he tried to apprise the battalion commander of what was happening.[8] With communications jammed, all the tanks headed north in Peeples's wake.

Peeples sighted five burning Army trucks south of the Euphrates bridge, with survivors returning fire. Deployed along both sides of the road, the tanks opened fire with coaxial machine guns. MPAT rounds fired from the main guns simply disintegrated some of the buildings, as well as a T-55 tank, anti-aircraft guns, and artillery pieces.

The decision to cut the fuel bladders loose came back to haunt Peeples as the tanks ran low on fuel. His problems were compounded by Army survivors huddled behind his tank, but an AAV appeared and they were hustled inside despite objections by Army Warrant Officer Marc Nash who wanted to go back to rescue his soldiers.[9] Unable to call in artillery over the clogged battalion net, Captain Jim Hawkins, the FAC riding in the loader's position on Dyer's tank, called in an air strike. The aircraft reported hundreds of enemy troops invisible from the ground and at least one T-55 tank. Battalion chose that moment to break in and abort the air attack because artillery support was belatedly on its way; the tankers seethed at yet another example of management from the rear. For their part the infantry were astounded that all the tanks had retreated to refuel.

The havoc wrecked the carefully rehearsed two-pronged attack plan. The regimental CO arrived on the scene and assessed the situation. The 1st Marine Division was breathing down his neck, anxious to cross the bridge. The overall commander on the scene accelerated the attack, making a controversial decision to send B Company across the Euphrates bridge to assume the tank company positions.

Unaware of the change, Peeples and his tanks were in no hurry to return. Worse, the fuel tanker had only 1,000 of the 5,500 gallons needed by the tanks. Every one of the electric fuel transfer pumps malfunctioned, so that fuel had to be transferred by hand. Dyer interrogated some of the Army survivors, who told him it was poor terrain for tanks, with swampy ground and numerous berms outside the town, and that the town contained bunkers built into the structures.

LtCol Eddie Ray, CO of 2nd LAR offered his Alpha Company to help with the attack into the city, but Col Ron Bailey (RCT-2) was forced to decline the offer in the face of objections from Col Joe Dowdy, the CO of RCT-1, who controlled Ray's battalion. The snap decision would come back to haunt Dowdy.

At about 1100hours 1/2 started into the city without tank support, but the advance faltered when the AAVs topped the railway overpass and encountered enemy tanks. The CAAT, infantry's anti-tank weapons, and Cobra gun ships destroyed five T-55 tanks, but the infantry was unwilling to advance without tank support. GySgt Randy Howard's 2nd Platoon raced back north, and at 1245hours led the attack across the river bridge, followed by CAAT, Bravo Company, and the battalion's Forward Command Element.

On the northern bridge ramp the Marines were isolated amid swarming *fedayeen*, and engaged in a close-quarters battle in which the muzzle blast from the tanks' 120mm guns was nearly as dangerous as the enemy. One hapless "technical" emerged from an alleyway only to be obliterated at close range by a 120mm round from Howard's tank *Lucille 2*.

The attackers veered to the east, into narrow streets where tanks scraped against the buildings and snarls of overhead power lines were a constant hazard. The vehicles tried to cross what seemed to be an open area about 100 meters square. It proved to be a sort of waste outfall, where raw sewage mixed with oil formed a deep bog covered with a thin crust of dried mud. Three tanks, two AAVs (including the AAV-C7 command vehicle), and several Humvees foundered in the stinking mess, the tanks sinking up to their engine decks. Behind them the Humvees of the CAAT were trapped in the narrow alleyways.

Men dismounted to hook up tow cables, only to flounder in the slime. SSgt Aaron Harrell dug desperately with an entrenching tool and his hands to locate the towing points on GySgt George Insko's tank, but mud flowed back as quickly as it was dug out. Towing cables and hooks broke. Once freed, vehicles lurched sideways and sank again. The Marines dubbed the place the shitbog.

In both Korea and Vietnam the tankers had quickly learned that VTRs were often most needed up front, but the two M88s and four Humvees of the maintenance group had been relegated to the rear of the regimental train. GySgt Wright and SSgt Cooke were trying to rush forward through the heavy traffic, but precious time was slipping away.

The infantrymen of B Company were struggling in a maze of narrow alleys. Behind them *fedayeen* kept appearing to fire at crews trying to recover the stuck vehicles. Communications, particularly the powerful radios in the command tracks, were disabled when the long antennae struck dangling power lines, zeroing out the electronic encryption sets.[10] The fight was degenerating into precisely the kind of confused brawl everyone feared.

Alpha Company followed in the wake of the first units, but plowed into a fire sack at the north end of the bridge. Defenders rushed the column and the grunts fired from the open AAV roof hatches. *Fedayeen* tried to board the AAVs, only to be cut down. In the close quarters fighting a track simply ran over an RPG gunner. One track was hit; ammunition stored inside ignited, the vehicle stalled in the street, and AAVs behind stopped. The thinly-armored tracks sought shelter by moving as close to walls and buildings as possible, while the infantry dismounted under a hail of small arms and RPG fire. Most of the company was now spread out along the first 500 meters of Ambush Alley, with some as far north as the canal bridge. Defenders kept rushing out of doorways and alleys, trying to close with the Marines.

In the confusion C Company had been relegated to last place in the battalion column, not first as planned. Charlie Company's tracks barreled through Alpha's fight, straight up Ambush Alley, and over the canal bridge, where they too were halted by intense fire coming from behind earthen berms.

Back at the refueling point Peeples was still unaware of the decision to accelerate and radically modify the attack plan. The crews of the Headquarters tanks were absorbed in pumping fuel and 1st Platoon was dropping a cumbersome bulky mine plow when Major Tuggle, the infantry battalion XO, appeared in a track, screaming for support. Alpha Company was clearly in trouble, but even Tuggle was unaware of C Company's plight or even its position.

Dyer made a snap decision and ordered Cubas's 3rd Platoon (really only Cubas and Kemper) north; Peeples was startled when the tanks raced away, but the sight indicated someone was in deep trouble. He ordered the balance of the tanks to break off refueling and head north as soon as possible, but his own tank suffered a mechanical malfunction forcing him to shift to another tank, adding to the confusion.

By now a running battle was taking place for several kilometers along the highway. Dyer's tanks had just taken up position atop the railway overpass when a T-55 tank appeared. Four enemy tanks were destroyed, but when Hawkins loaded an MPAT round, the casing refused to seat. No amount of hammering would drive it into the chamber. Finally a crewman dismounted to clear the jammed round with the rammer staff, and Dyer charged into the fight at the north end of the river bridge.

The Forward CP, trapped with B Company at the shitbog, had only a vague idea of the rest of the battalion's positions. Grabowski had reported the capture of the canal bridge to regiment, but between the chatter over the radios and the poor reception, the Forward CP failed to receive the message.

North of the canal C Company was taking heavy punishment. Charlie

Company had no FAC because it was supposed to bring up the rear, and the most advanced FAC was with B Company. This would lead to the greatest tragedy of this confused battle. Battalion had issued a written order that no Type 3 Close Air Support would be employed without the express approval of the CO, but the system failed.[11]

Bravo Company's FAC could see vehicles—the trailing elements of C Company—and aborted air attacks north of his position. Assured that there were no friendlies north of the canal, he freed the air support.[12]

Marker rockets fired by two A-10s were invisible to B Company's FAC; the A-10 pilots mistook AAVs (vehicles unfamiliar to the pilots) for Iraqi vehicles forward of B Company positions; B Company's FAC and CO were still unaware of any Marines north of the canal.

In fact C Company was fighting for its life, unable to communicate with the artillery or 81mm mortars. One of the units badly hit was the company's 60mm mortar section. The AAVs were drawn up mostly along the western side of the highway, facing the heaviest fire. One AAV, C-205, took on a load of wounded and broke free, heading south. All this commotion did not escape the A-10s overhead.

Just as Marines piled into other AAVs, the two A-10s rolled in for repeated cannon and missile runs. As C-206 crested the canal bridge a Maverick missile impacted just behind it; the damaged rear ramp dropped, spilling men into the roadway. The vehicle kept going, dragging the ramp. C-208 took a direct hit from a Maverick, lifting the vehicle several feet off the concrete and blowing bits of vehicle and passengers into the air. The driver and the vehicle commander survived, sheltered by the massive engine and flimsy internal partitions.

AAVs C-201 and C-203 were hit by cannon fire. Red emergency flares fired from the ground did not dissuade the pilots. C-201, in flames and with steering disabled, crashed into a building. The grunts spilled out and went to ground in nearby buildings, but in their haste salvaged limited ammunition, and no water. Their only salvaged radio was rendered useless by dead batteries. Many of the men took shelter on the roof of a building, soon called The Alamo, on the east side of the broad boulevard, near the south end of the canal bridge.

The B Company FAC was finally able to abort further attacks, and the puzzled pilots flew away.

The disaster had left less than a platoon of effectives north of the canal bridge.

Southwest of the city parts of V Corps and the 1st Marine Division were caught in a massive traffic snarl along Route 1. Seth Folsom and the rest of 1st LAR could clearly see the pillar of black smoke rising from the city, but were unaware of the magnitude of the fighting.[13]

At about 1400 hours Peeples, with four tanks and a gaggle of AAVs, arrived at A Company's position and dismounted to ask what was going on, ordering Swain and Thompson farther up the road. Those tanks took up positions farther north with Alpha Company, blasting away at positions around the small bridgehead. The tanks gave the hard-pressed infantry some respite, as most of the fire was now directed at the tanks. The tanks' heavy machine guns and cannon soon began to suppress the incoming fire. A helicopter pilot overhead reported "waves" of Iraqis trying to rush the immobile vehicles.

One of the two surviving Charlie Company tracks, C-206, limped into the Alpha Company bridgehead, trailing a shower of sparks from the open ramp as it dragged along the pavement. Just as it reached relative safety an RPG entered the open hatch, and it was gutted by an ammunition explosion. C-210 veered around it and stopped in the street. A squad of C Company grunts, mistakenly embarked in the confusion north of the canal, spilled out.

At the north end of the river bridge Dyer, exposed in the TC's hatch, was blasting away at enemy positions while Hawkins leaned far out of the loader's hatch to coordinate air support. In the desperate fight Hawkins had to keep numerous balls in the air: coordinating air support, loading the main gun, and firing the roof-mounted machine gun in front of his hatch. Rounds fired by a skillfully hidden sniper were snapping around their ears. As bad as things were for the beleaguered A Company grunts, they were now the only intact part of the battalion.

One C Company survivor, Sgt William Schaefer, tried unsuccessfully to solicit help from infantry officers who were still under instructions to hold at the north end of the river bridge until relieved by 8th Marines. The truck-mounted 2/8 dared not try to run the gauntlet of fire. The Marines dismounted to clear the enemy from both sides of the highway, working north in a broad wedge, and were still entangled in fighting near the railway overpass. Schaefer finally ran to Peeples's tank to explain the disaster north of the canal.

At about the same time, Dyer received a disturbing radio message from the leader of C Company's 3rd Platoon: "This is Palehorse 3. We need assistance. We need tank support."[14] Normally platoon leaders do not transmit on the battalion's primary radio net; Dyer realized things must be seriously wrong at the canal bridge. Just then he spotted a muzzle flash in the deep shadows of a window, and blew the building away. The sniper fire stopped, but Hawkins was crumpled in the floor of the turret. Dyer thought he was killed by the unexpected muzzle blast, but Hawkins stirred and gave Dyer a dark look.

Peeples felt he had to go to C Company's aid; he decided to take two tanks north and leave two—Thompson and Swain—to help A Company. Schaefer organized a relief force, and would follow in an AAV. Peeples and Dyer raced three kilometers up Ambush Alley at over 70 kilometers an hour, soon leaving the slower AAV to its own devices. As the tanks blew through the smoke and sparks of RPG rounds, Peeples caught sight of an air recognition panel draped over the roof parapet at the Alamo. The tanks did not have time to investigate.

To the south F/2/8, closest to the road, was ordered to remount its trucks and blast through to the positions north of the river, despite the risk. Filipowski, his tank repaired at last, had struggled through the massive traffic jams only to be appropriated by staff officers who wanted the tank to escort supply trucks along the still dangerous highway. He could only listen to the radio traffic of the fight.[15]

Enemy fire continued to savage C Company. One of the remaining amtracs was struck and set afire by an RPG, and Navy Hospitalman Louis Fonseca Jr. ran through the small arms and RPG fire to the vehicle. Helping five wounded into the medical track, he provided what aid he could for the wounded, including two men with severed legs. When another RPG disabled the medical track, Fonseca grabbed nearby Marines to move four of the wounded, and slung the fifth over his shoulders for a mad dash through intense fire to another vehicle. Fonseca was awarded the Navy Cross for his actions.[16]

Peeples's two tanks topped the canal bridge, and raced into C Company's position; the command track, with its distinctive diamond symbol, was parked near the foot of the bridge ramp. The company commander jumped onto Peeples's tank to direct fire, while Dyer continued another 500 meters to the north to relieve an isolated platoon. Hawkins was now able to bring in several flights of Cobra gunships, and the Iraqi fire began to die down a bit. After firing a few rounds, Peeples took his tank back into Ambush Alley to investigate the air recognition panel.

The Alamo had by now held out for over two hours, and the grunts had sallied out to salvage what they could from the disabled AAVs: ammunition, a bit of water, and a few drained batteries that allowed them to communicate by clicking once for "yes," twice for "no." Peeples halted his tank and dismounted to confer with the grunts and help look for lost Marines. He loaded the wounded on the engine deck, but the rest declined his suggestion that they walk out under the cover of his tank.

The two reinforcing tanks from 3rd Platoon, Captain Cubas and GySgt Kamper, had also blundered into the shitbog. Kamper's driver tried to tell him that he could not see, but Kamper thought the periscopes were obscured. Now the dust-clogged NBC filter caught fire, filling the interior with smoke.[17] As the crew dismounted an Iraqi rushed out of a nearby building. The gunner, Sgt August Nienaber, tossed his 9mm pistol to the driver. LCpl Joshuah Mouser caught it, and emptied both his and Nienaber's pistols at the enemy. Unhurt, the stunned Iraqi dropped his own AK-47 and ran for it.

Frustration continued to mount at the shitbog. GySgt Howard improvised tow cables, and managed to free both Insko's and Dillon's tanks from the mire. About 1600 the two VTRs arrived, led through the fighting and chaos by Sgt John Ethington, who had dismounted to lead the vehicles on foot. One sank into a hole, and had to be pulled out by the other. Efforts to extricate Howard's tank led to the sinking of an M88. It was pulled free by a tank, but the other M88 foundered while trying to recover the AAV-C7 command track. At the end of the festivities, these two remained stuck in the mire with darkness approaching.

Apprehensive of what could happen to small, scattered perimeters LtCol Rick Grabowski was trying to consolidate his battalion. He ordered the vehicles at the shitbog stripped and abandoned. A platoon of infantry, a few men from the command group, and a small detachment from CAAT would watch over them while the balance of B Company moved up

Ambush Alley to C Company's position north of the canal.

Alpha Company, still hard-pressed, formed up unto a convoy of Humvees and AAVs and started up Ambush Alley. The convoy sighted the panel that had drawn Peeples's eyes to the Alamo, and hurriedly dropped off what supplies were at hand: a canteen (empty) and some radio batteries (the wrong size).

Alpha Company arrived north of the canal, and parts of C Company and Humvees from CAAT returned to rescue the group at The Alamo. Dyer, still the only heavy firepower at the far northern point of advance, established a roadblock to intercept enemy forces coming from the north. Cubas and Kamper arrived amid the usual flurry of RPGs to find Grabowski's command group at the south end of the canal bridge. The consolidated group moved north, pausing only to recover bodies and disable the radios in the abandoned AAVs. The Marines were at last beginning to establish a series of defensible positions for the coming night.

While all the mayhem was taking place in the city, 3/2 had seized the uncompleted Route 1 bridge west of the city with little opposition.

The most ferocious fighting in an-Nasiriyah died with the light, but the situation remained confused. With communications disrupted, losses in 1/2 were feared heavier than they were. Eddie Ray again offered the assistance of 2nd LAR; Alpha Company was sent through to support 1/2, with a platoon detached to support 2/8 south of the city. After nightfall a small party of British Commandos passed through the town headed west, but returned after being ambushed before reaching 3/2's position at the Route 1 bridge. When 3rd LAI attempted to move up Route 1 north of the river they encountered a well-laid U-shaped ambush which was fortunately triggered prematurely. The battalion was able to disengage under the protection of air strikes.[18]

The Marines at the isolated northern perimeter were exhausted. Dyer was babbling nonsense. Peeples fell asleep while climbing aboard his tank and slept on the engine deck.[19] That night a roadblock north of the canal flagged down an Iraqi ambulance with lights flashing. It proved to be carrying the CO of the Iraqi *23rd Infantry Brigade*, disguised as a paramedic with his uniform, documents, and a large amount of US currency stashed in plastic bags.

All through the night the Marines in the exposed positions held the enemy at bay with sniper and cannon fire. To the south the staff and supply personnel labored to organize a resumed attack and a supply convoy to break through to Grabowski at first light. In a gamble, Major Tuggle sent AAVs with supplies racing down Ambush Alley in the pre-dawn hours. They were unopposed, the enemy still asleep.

Dave Banning's Alpha/1st Tank was caught up in the enormous traffic jam generated by the battle. "We really needed fuel. They sent a refueling element back to us an hour later and I tried to figure out what was going on. The detachment commander for the refueling unit didn't know what was going on either. As soon as we refueled, we pulled into the tail end of 1st Marines and they had pulled off the main MSR. The main MSR kind of bypasses Nasiriyah to the south. They were on the MSR that headed right to Nasiriyah. The entire regiment was parked along this two-lane highway and no one knows anything. The entire unit is just stopped. The 1st Marines have nearly 1,000 vehicles on the road and they're just stopped. We sit there for a bit trying to figure out what's going on and it's starting to get later in the evening, so we decide to move up. We start going down the middle of the road and everyone parts the way for us. As we're moving up there, everyone is clapping and taking pictures and cheering like we're a bunch of rock stars. We get up to the regimental CP and it turns out that Task Force TARAWA had been engaged in a pretty stiff fight."

At a halt Banning tried to sleep on the engine deck, but "I was told to get to the regimental CP. I went back up there and Colonel Dowdy took me and we went up to talk to Colonel Ron Bailey who was the commanding officer of the infantry regiment with Task Force TARAWA. We were trying to get a feel of what was going on. The Task Force TARAWA guys looked like they had just had their asses kicked. Everyone was walking around with their chinstraps undone and they just looked like they had been hit hard. These two colonels were trying to figure out what we were going to do. I remember Colonel Bailey telling Colonel Dowdy that they needed to figure out what the MEF wanted all of us to do. I can't imagine the stress they must have been under. The picture we had of what was going on in Nasiriyah was that we knew there were friendlies along the T intersection in the north, friendlies along the south side of the river and nothing in the four-kilometer stretch in between."[20]

At about 0430hours on 24 March the Marines began realigning units for the passage of 1st Marine Division. The 3/2 began to withdraw back down Route 1 from the western bridge to clear that route for the passage of the 5th and 7th Marines. At first light Cobra gunships began to roam the battlefield, shooting up taxis that the *fedayeen* used as transports. North of the canal 1/2 expanded its area of control west to the T-intersection where Route 7 turns north from the city, one of its original objectives. Eddie Ray began to pull back his scattered force in preparation for moving through the city. The 1st Marine Division planned to barge down Ambush Alley with its combat components, but divert the vulnerable logistical trains down the south shore and across the Route 1 bridge west of the city.

During the night Tuggle had organized a large relief convoy with whatever resources were at hand, throwing in Peeples's repaired tank, the late-arriving Filipowski, and a platoon of LAVs for good measure. The 155mm howitzers of 1/10 swept the convoy's route with a rolling barrage to discourage interference.

The staff officers were still debating their plan for passage of RCT-1when Tuggle's relief convoy rolled through Ambush Alley. Fox Company 2/8 and a gaggle of staff officers watching from the southern end of the river bridge came under fire from the main hospital compound on the north bank. The Marines were calling in artillery fire when they learned that prisoners from the 507th Maintenance Company might be held there. Suddenly Iraqis started surrendering; a stream of *fedayeen* began to flow west out of the hospital grounds while a doctor emerged to organize the surrender of the facility.

The progressive collapse of *fedayeen* resistance caused 1st Marine Division to quickly alter the plan for passage through the city. Second LAR Battalion and tanks led the advance of the 1st RCT through Ambush Alley and on toward Baghdad, with logistical trains following.

Banning: "My second platoon was still with 2nd LAR. 3/1 is told to set up a picket line between the north and the south bridge and provide security for everybody else going through. My company was going to lead 3/1 into this town. We were going to go in and make the initial entry at night. If we faced any resistance, we were supposed to whack it and make things permissive enough for amphibious assault vehicles (AAVs) to come in, dismount the grunts and provide the close-in security. Around midnight, I took my company-minus and we rolled into town. We went into this built-up area with no infantry support. I personally didn't have a significant problem with it because as you looked at the map you could see the way the town was laid out and there was a little bit of space between the buildings and the road, a little bit of standoff. I was confident that if we could get in there quick enough, we'd be able to destroy any strongpoints before they could mount any kind of coherent defense against us."[21]

By early evening RCT-1 had reached the western T-intersection west of the town, taking fire from both sides of the highway and replying with 25mm cannon fire and artillery.

Task Force TARAWA retained responsibility for the city itself, but with the change in plans was ordered to secure the west side of Route 7 between the railway line and the north bank of the river.

At about 2000hours the 1st Marine Division's lead elements began crossing into the city, amid a huge snarl over who had priority of movement through the chokepoints. The lead battalion of the 1st Marines with two platoons of tanks attached moved across the river bridge and into Ambush Alley to establish a picket line. The tanks peeled off to secure key intersections.

Banning: "So we got in there, it was the middle of the night, and the Iraqis didn't seem too interested in putting up any type of coherent defense. This company is now spread out along four kilometers stretched out through the town and actually being in the town, with the buildings in closer proximity to the road than they had appeared on the map, making me a little nervous. There was a lot of space in between tanks, not really mutually supporting each other anymore. . . . There was a gap in 3/1's infantry companies as they spread out and, as a result, my company XO and I didn't have any infantry close to either of our tanks. We spent all night watching each others' backs and when the sun came up, 3/1 pushed in the third rifle company and filled in the middle of the line. I felt a lot better with a squad of grunts close to the vehicle. Our tank infantry coordination was a little ad-hoc. I basically hopped off the tank, grabbed the closest squad leader and gave him a quick lesson on how to use the grunt phone on the back of the tank and how to tell if we were getting ready to move the vehicle. As the sun comes up, the Iraqis figure out we're there and start shooting back. Things kind of wax and wane during the course of the day, with the heavier fighting occurring down towards the southern end of the town.

There were some pretty good firefights going on. The battalions were just trying to get through there as fast as they could. I had a concern with the fact that there was a pretty significant time gap between one battalion leaving and the next battalion coming in there. We were beginning to run out of fuel and weren't going to be able to sit there for the rest of the day while these people dilly-dallied around outside the town. I wanted them to get going. The rest of the regiment finally ended up getting through the town and I pulled the company out to the north."[22]

Fedayeen holdouts were still active and 3/1 began to lay down suppressive fire on its way through the city, firing mostly to the east. At about 0330hours Grabowski, at the T-intersection north of the canal bridge, grew alarmed at all the firing as 3/1 approached. He had his men mark positions with chemical light sticks, and walked out into the highway to block the convoy and stop the random firing.[23]

Charlie Company, 2nd LAR searched a broad arc east of the city, then about noon followed 3/1 when that battalion broke down its picket line and followed its parent regiment north along Route 7.

West of the city movement along Route 1 slowed to a crawl when the lead battalion, 3/5, encountered heavy resistance. Attempts by 1st LAR to move off the road into farmland to the east ended when it became apparent that the high paddy dikes with limited gaps restricted movement. By late evening the battalion was still drawn up behind 3/5. Tanks—with superior night vision devices—were called upon to investigate movement in the broken ground to the east but saw nothing; the artillery saturated the area with HE and DPICM.[24]

As the division moved north TARAWA stayed behind to secure the still hostile city and protect lines of supply and communication. They scoured numerous Iraqi military compounds, and searched the hospital compound but found no American POWs. The Marines also recovered remains of those killed in the 507th Maintenance Company disaster, and the bodies of dead Marines from unmarked graves. In keeping with Islamic beliefs, most had been quickly buried in graves dug in back gardens.

By 26 March platoons of LAVs from C/2nd LAR were still required to escort convoys down Ambush Alley. Civilians, with *fedayeen* intermixed, were fleeing the fighting across the western bridge, so Peeples's 3rd Platoon assisted A/1/2 by using a mine plow to pile wrecked cars to bar passage.

Following an unsuccessful night attack on 2/8's positions, in which Marine units inadvertently fired on each other, 27 March was spent sweeping the sprawling old part of the city west of Ambush Alley. On 28 March Ambush Alley was opened to unescorted convoys, though the western districts were still under *fedayeen* control. The task force distributed flour confiscated from Iraqi Army warehouses to starving civilians.

Local residents continued to relay rumors of American prisoners being held in the city, and on 29 March the Marines received credible intelligence of a female prisoner at the Saddam Hospital in the unsecured part of the city. Asked to provide tank support for a high-priority mission, Peeples selected Cubas, Kamper, and Filipowski. The tanks were sent to the western bridge to support Task Force 20, composed of Navy SEALs and Army Rangers.

Around midnight, under cover of a noisy diversion of artillery fire and air strikes, helicopters swooped into the hospital compound from the west and north and spirited Jessica Lynch, a much-celebrated survivor from the 507th, to safety. Far from the spotlight, Kamper used the mine plow on his tank to smash aside the blockage he had created on the western bridge, and the tanks followed Task Force 20 three kilometers through the city to the hospital. The tanks and Rangers secured the hospital grounds, and conducted a detailed search for any other prisoners, or remains of any dead.

On 3 April Task Force TARAWA finally left the city, and began to patrol the line of supply through southern Iraq.

NOTES

1 Reynolds, *Basrah, Baghdad, and Beyond*, p. 79.

2 Ibid,p. 74.

3 The CAAT was a highly mobile unit mounted in lightly armored Humvees, and armed with a diverse mix of TOW missiles, light and heavy machine guns, and Mk19 40mm automatic grenade launchers. Their job was basically to look for trouble. Marine Corps tank platoons consist of five vehicles, and can be tactically subdivided into a light section of two tanks under the senior NCO and a heavy section of three tanks under the platoon leader.

4 Pritchard, *Ambush Alley*, p. 24.

5 The general outline of the following section is taken from Glover, H., *Command Chronology Report for Calendar Year 2003, Company A, 8th Tank Battalion*; Reynolds,

Basrah, Baghdad, and Beyond, p. 75–85, and Pritchard, *Ambush Alley*. Where significant discrepancies exist, the Glover account is given priority. Some sources indicate the 507th Maintenance Company convoy had eighteen trucks; Richard S. Lowry, *Tanks in the Garden of Eden*.

6 Driven by PFC Lori Piestewa (died of injuries in captivity), was carrying Master Sergeant Robert Dowdy (killed instantly) and PFC Jessica Lynch (injured and captured). Far to the rear, public affairs officials and the press made much of Lynch's supposed heroics, actually the deeds of PFC Patrick Miller (captured). To her credit, Lynch consistently strove to disabuse others of any notions of her supposed heroics. The lead element of the convoy acted according to doctrine, that any mobile vehicles should move out of the ambush kill zone, then attempt to assault the ambushers from outside the kill zone. Of course the 507th was not specifically trained or equipped to assault the ambush, and there were in any case too few of them to take on hundreds of ambushers.

7 Dunfee, *Ambush Alley Revisited*, p. 44. Dunfee strongly disputes the account of the battle presented by Bing and West, *The March Up*.

8 Pritchard, *Ambush Alley*, p. 31. The tank unit commanders thought the battalion staff tried to micromanage the action, and there was a constant radio chatter that often precluded critical messages from getting through.

9 Lowry, *Marines in the Garden of Eden*, p. 145–147.

10 Dunfee, *Ambush Alley Revisited*, p. 45.

11 Type 1 and Type 2 CAS are directly controlled by ground observers. Under Type 1 CAS the target is clearly visible to the observer, under Type 2 the target is known but not directly visible. Type 3 grants pilots the discretion to attack any target within a geographic area, turning it into a free fire zone. The Marine Corps seldom utilizes Type 3.

12 Lowry, *Marines in the Garden of Eden*, p.202–203.

13 Folsom, *The Highway War*, p. 120.

14 As quoted in Pritchard, *Ambush Alley*, p. 264.

15 Lowry, *Marines in the Garden of Eden*, p. 212.

16 Raymond L. Applewhite and Eric Schwab, untitled news release, unpaginated.

17 Lowry, *Marines in the Garden of Eden*, p. 199–201. NBC: Nuclear, Biological, and Chemical Warfare. The charcoal filter is designed to trap particulates, finely-dispersed fluid droplets, and large-molecule chemicals before air is blown into the crew space.

18 Folsom, *The Highway War*, p.134.

19 Lowry, *Marines in the Garden of Eden*, p. 269–273, 281.

20 Lessard, *Interview with Major Dave Banning*, p. 10.

21 Ibid, p. 11.

22 Ibid, p. 11.

23 Lowry, *Marines in the Garden of Eden*, p.314–316.

24 Folsom, *The Highway War*, p.134–142.

The old M60A1was the mainstay of the Corps tank program through most of the post-Vietnam era. This one is patrolling near the Beirut International Airport in April, 1983, prior to the October suicide bombing. —*Defense Imagery*

The M60A1 tanks initially deployed to Saudi Arabia arriver without reactive armor and painted in "European" camouflage. Note the hasty re-painting of some vehicles, and the sandbags as additional protection for the driver.—*Sgt D.R. Renner, Defense Imagery*

An M60A1 of Delta Company, 2nd Tank Battalion fitted with reactive armor and an M9 dozer kit rehearses breaching operations for Operation DESERT STORM, late January 1991. —*SSgt M. A. Masters, Defense Imagery*

Saudi troops examine an LAV-AT of H&S Company, 1st LAI. This is the type of vehicle destroyed in the friendly fire incident during the fighting for al-Khafji. An LAV-25 is in the background. —*Cpl D. Haynes, Defense Imagery*

The more modern M1A1 equipped only select companies of the tank battalions in Operation DESERT STORM. This one, equipped with a mine plow, is passing a wrecked Iraqi Army truck in Kuwait, 24 February 1991. —*SSgt M. A. Masters, Defense Imagery*

The Iraqi tank forces quickly proved no match for the better-trained American forces. This T-55 was abandoned intact in Kuwait, 1 February 1991. —*Defense Imagery*

Iraqi anti-aircraft guns like this abandoned ZSU-23-4 (four 23mm radar-controlled guns on a tracked chassis) were priority targets because of the threat they posed to low-flying aircraft in a chaotic battle. —*Defense Imagery*

The feared T-72 proved to be a paper tiger. Its main gun was slow to reload, and it often exploded spectacularly when hit. Note that on this one the turret has been lifted up and back by an internal explosion. —*TSgt Joe Coleman, Defense Imagery*

M60A1 tanks, Humvees, and an M88 recovery vehicle (background) pause near a burning oil well in Kuwait, 27 February 1991. —*CWO2 Ed Bailey, Defense Imagery*

PROVIDE COMFORT was the effort to secure the Kurdish region of northern Iraq against Saddam Hussein's retaliation. The Marines deployed to the region were supported by air-portable LAVs. —*PH3 J. R. Klein, Defense Inagery*

An M1A2, probably from 1st Tank Battalion, maneuvers while passing through an-Nasiriyah, 25 March 2003. The author has located no photos of Alpha Company, 8th Tank Battalion, from this battle. —*Defense Imagery*

A TOW-equipped Humvee of the CAAT, Task Force TARAWA, maneuvers during the final fighting for an-Nasiriyah, 1 April 2003. —*Defense Imagery*

Condition of the main highway bridge west of an-Nasiriyah was the cause of considerable concern to planners. It was the desire to bypass this bridge that in part led to the fight for the city. —*Defense Imagery*

East meets west; a heavily-loaded donkey cart passes a heavily loaded Marine Corps tank in the streets of Baghdad, 9 April 2003. —*Sgy P. L. Anstine II, Defense Imagery*

A tank from 1st Tank Battalion with grunts from the 3rd Battalion, 7th Marines in the streets of Baghdad, 9 April 2003. —*Sgt P. L. Anstine II, Defense Imagery*

A burned-out Marine tank sits beside the highway somewhere south of Baghdad, 3 April 2003. —*LCpl A. A. Plaza, Defense Imagery*

An M1A1 of Charlie Company, 1st Tank Battalion crosses an overpass along Highway 6 en route to the tank assembly area at the Olympic Stadium in Baghdad, 12 April 2003. —*Sgt P. L. Anstine II, Defense Imagery*

A tank supports 1st Battalion, 8th Marines in the streets of Fallujah, 9 November 2004. —*LCpl J. A. Chaverri, Defense Imagery*

A Charlie Company tank merges into the chaotic civilian traffic on Highway 8 between Baghdad and Kerbala, 18 April 2003. —*Sgt P. L. Anstine II, Defense Imagery*

A tank attached to 11th MEU in a raid on the Muqtada Militia Headquarters in al-Najaf, 24 August 2004. Note the absence of extra gear on this tank operating from a base area. —*GySgt D. J. Fosco, Defense Imagery*

A tank from Charlie Company, 1st Tank Battalion in the streets of Baghdad, 14 April 2003. —*Sgt P. L. Anstine II, Defense Imagery*

This tank has broken down the side of a road and slid into an irrigation canal outside Fallujah. Several AAVs are hooked to the tank by tow bars in an effort to extricate the tank. The deep ditches, soft ground, and poor narrow roads made this a constant hazard. —*Bill Hayes*

A tank-infantry patrol in the streets of Haditha, 27 January 2005. —*GySgt K. W. Williams, Defense Imagery*

An engineer-armored bulldozer flattens a building in Fallujah. These armored dozers proved invaluable in reducing enemy positions and clearing rubble and obstacles in an urban battle. —*Hayes*

An LAV-25 of 3rd LAR Battalion patrols near al-Rutbah, 1 January 2005. —*LCpl R. A. Hilario, Defense Imagery*

Dismounts from Delta Company, 2nd LAR Battalion search an Iraqi farm near Balad, 4 August 2008. The dense vegetation of the farmlands in the river valleys contrasts sharply with the surrounding desert. —*LCpl A. L. Hunt, Defense Imagery*

Marines of Delta Company, 3rd LAR Battalion repair a disable LAV-25 on the road in Anbar Province, 17 April 2009. —*LCpl B. A. Kinney, Defense Imagery*

Marines and Navy corpsmen work to free an LAV-25 bogged in soft sand, Nineveh Province, 13 May 2009. —*LCpl B. A. Kinnney, Defense Imagery*

An LAV-25 attached to 12th MEU(SOC) patrols the Kandahar International Airport. The air-portable LAVs were the first Coalition armor to arrive in Afghanistan; photo taken 15 January 2002. —*Capt C. G. Grow, Defense Imagery*

A heavily-loaded LAV-25 from Alpha Company 2nd LAI, attached to 26th MEU(SOC) patrols in an Afghan village near the airport, 17 January 2002. Scouts are standing in the rear hatches, and there are three locals on board as guides and interpreters. —*Capt C. G. Grow, Defense Imagery*

M88A2 heavy retrievers were brought in as utility vehicles before tanks. This one is helping repair an ABV engineer vehicle in Helmand Province, 20 March 2011. —*Cpl John McCall, Defense Imagery*

Scouts from Charlie Company, 3rd LAR search for a Taliban mortar observer, Helmand Province, 16 March 2011. —*Sgt Jeremy Ross, Defense Imagery*

The Assault Breacher Vehicle (ABV) is an engineer asset based on the M1 tank chassis. This vehicle from 1st Combat Engineer Battalion is operating in support of 3rd LAR, Helmand Province, 15 March 2011. —*Cpl John McCall, Defense Imagery*

In the latter stages of both Iraq and Afghanistan all types of units increasingly utilized Mine-Resistant Ambush-Protected heavy trucks (MRAPs) like these in use by 1st Battalion, 5th Marines in Afghanistan, 3 November 2009. —*LCpl James Purschwitz, Defense Imagery*

This M1A1 from Alpha Company, 2nd Tank Battalion, is fitted with a combination mine plow and rake. Helmand Province, 11 August 2011. —*Cpl Marco Sancha, Defense Imagery*

CHAPTER SEVEN

The Low Road to Baghdad

"Speed is of the essence in this endeavor."
—LtCol Mike Oehl, 2nd Marine Tank Battalion[1]

A S THE 1ST MARINE Division struggled to cross the Euphrates on the evening of 24 March the 1st Tank Battalion suffered one of the non-combat losses that the volatile mix of numbing fatigue, poor conditions, and powerful machinery seem to make inevitable.[2]

Charlie Company tanks were ordered to halt on the southern approaches to the uncompleted Route 1 Bridge and await instructions. Some had just completed an arduous refueling, squeezing fuel from the huge rubber bladders into the internal tanks. Many crewmen dozed off during the long halt. The tanks were under pressure to pick up the pace, so at 0100hours the company was given the order to move, only to halt on the bridge when the TOW platoon reported it was unable to reconnoiter a clear route on the far shore. Tanks C-23 and C-24 did not hear a radio message to move to the side of the bridge to clear a traffic path and the CO, Captain Brendan Rodden, walked forward and relayed the order.

Just after 0300hours new orders came; the drowsy drivers cranked the engines, and the company moved out. Captain Rodden lost sight of the tank ahead, and instructed his driver to speed up to close the gap. The column sped on through the night, each tank following the marker lights of the tank ahead. An hour later Rodden discovered that he was following the wrong tank. Repeated radio calls failed to contact the missing tank, *Harvester of Sorrows*. Rodden hoped that the tank had only gotten lost in the darkness, but the tank and its crew had simply disappeared.

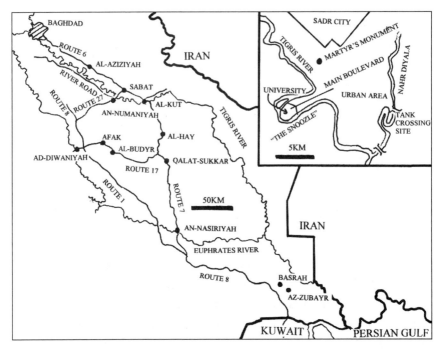

MAP 3. Marine Corps Area of Operations, Southern Iraq and Baghdad.

Five days later the missing tank was found submerged in the river below the bridge. Apparently when the driver moved to the right to clear the traffic lane, he drove through a gap where a sidewalk was under construction. The exposed reinforcing bar crumbled under the weight, and snagged the tank tracks as it tumbled over the side. The tank flipped over, striking the water upside down; no escape was possible. Four Marines died in the sunken tank, SSgt. Donald C. May, Jr., Corporal Robert Marcus Rodriguez, Lance Cpl. Patrick O'Day, and PFC Francisco Martinez Flores.[3]

North of an-Nasiriyah Route 7 pinches down to two lanes; the 1st Marines continued north along that highway, while 5th and 7th Marines swung west, following the main highway, Route 1. On 25 March the leading element, 2nd LAR, gingerly picked its way through a litter of wrecked and abandoned vehicles along Route 7, victims of Coalition air strikes. Behind were the 1st Marines, clogging the road with another hundred AAVs, a company of tanks, and innumerable trucks. The advance was further slowed by petty ambushes, but no one wants to be killed in a petty ambush;

each was laboriously cleared by tank fire or small infantry assaults.

The Route 1 axis of advance was clogged by 3/5 as they rooted out the survivors of the previous days' fighting. The battalion decided to launch a Dragon Eye to check out the road ahead, but when a helicopter MEDE-VAC's flight path passed through the area the little drone was deliberately crashed. It flew apart as designed, but the wind scattered the lightweight parts all over the desert—not as designed.[4]

Dan Hughes had been told he would be attached to 3/5 for the push toward Baghdad, under vague orders. "It was kind of like 'Okay, listen lieutenant, here's what I want you to do. You're gonna be in front of the battalion, and you're simply gonna drive down the highway.'" Colonel Mundy had assembled the officers and described the debacle in an-Nasiriyah: "That was the time when he said 'Make sure we got our heads on straight that we are in a *war*. We're not doin' a road march. This is a war.'"

The next day the task force approached a large airfield, and noticed a man wrapped in a blanket; they thought it odd, but kept moving. The lead tanks reported armed enemy, and were cleared to engage. "I could see his vehicle get hit by a couple of RPGs. The RPG hits out in front of my tank. It's right in the front, basically the loader's side of the turret. . . . We start engaging, and at that time we had this massive berm on the side of the road. They were essentially on the other side of the berm, shooting down on top of us." The tanks fired main gun rounds into the berm, and Hughes called for infantry. The CAAT team arrived and maneuvered their Humvees up and over the berm.

Flanked, the Iraqis were trapped. "Some of them tried to go over the berm and were engaged by my section of tanks with coax. Now they're kind of in trouble." As the CAAT team cleared the enemy, Mundy kept pushing for Hughes to move on up the highway.

About two kilometers south of a compound near the highway the task force's lead elements came under mortar fire, and returned fire with main gun and coax.

While fighting was still going on in the late afternoon Nature achieved what the Iraqi Army could not. While Hughes' tanks were waiting for additional infantry and air support, "That's when that dust storm from hell showed up. That kind of squashed all that."

The dust, whipped by 50mph (80kmph) winds, caused a "brownout."

The advance ground to a halt and the mechanized units went into huge defensive coils. Darkness brought a chilling hour-long rainstorm, starting at about 2130hours. The rain flushed dust from the air to fall as a rain of mud, followed by sleet and then hail. Mike Oehl was caught at the 5th Marines headquarters, 26km from his own headquarters. Unable to see anything in the blackness, Oehl navigated by GPS, and shouted instructions to his driver over the storm's din.

Massive traffic congestion along the limited road network was causing more problems than the collapsing Iraqi military. Captain Dave Bardorf was in charge of 2nd Tank's combat trains, and had not slept in days. Peering through night vision goggles he shouted instructions to his driver until the exasperated driver "shook my shoulder and told me we'd been stopped for five minutes." Bardorf collapsed into a stupor and slept through the night.[5]

The trying conditions of the "Orange Crush" demonstrated the utility of both design features and personal initiative. Benz: "Marine tanks have this device that blows the air backward for a couple of seconds to clean out the air filters. I'm obsessive compulsive. We had lots of toilet paper. I was a big Nazi about cleaning windshields. When you finally clean your windshield, you feel like, 'Wow, you can really see a lot better.' You see a lot of other units driving around with orange mud, then they turn on the windshield wipers and just smear it. 'Hey, asshole, you're just begging for an accident. How can you see the bad guys?' Stupid little things like we put pantyhose over all the air intakes. It's simple. It's discipline. It's give-a-shit factor."[6]

By the morning of 26 March the infantry were reduced to living like pigs in wallows. In the vehicles, dust and wind-driven rainwater had infiltrated through every crevice, filling everything with soupy mud. Weapons and ammunition had to be laboriously cleaned, part-by-part and round-by-round.

The advance up Route 1 continued to be an infantry slog, clearing numerous small ambushes in the advance toward the next objective, ad-Diwaniyah. The city itself was of no tactical value, but the military airfield at Hantush, north of the city, was needed as a Forward Armament and Refueling Point (FARP) for tactical aircraft. Near the air base Route 27 branched to the east and the city of al-Kut, on the 1st Marines' axis of advance. The overall plan was to feint toward Baghdad from ad-Diwaniyah, but for the entire MEF to shift east onto Route 6 and approach Baghdad from the east.

Sam Crabtree remembered the Field Trains as being generally safe, but he was lucky. The only ambush occurred while he was in the field repairing a tank. Second Lieutenant Aaron Klein was the Legal Adjutant for 1st Tank: "Out of nowhere we start taking small arms fire, and mortar rounds, and some RPGs." The enemy was behind a berm about 160 meters off the highway. The Marines suppressed the enemy fire "As we got back to the road, it started from the other side. Apparently they had been waiting for us, or realized we had halted, and set up some kind of ambush there." Eventually the battalion sent tanks and TOW carriers back to eliminate the enemy.[7]

Stress and fatigue are the constant enemies of those on night watch on a battlefield. About 0430hours the 2nd Platoon of Folsom's D/1st LAR reported movement to its front. Other platoons verified the observation, reporting men and vehicles, and figures setting up and firing a mortar in plain view. Folsom was puzzled. Why would the enemy be moving about so boldly? Finally the 81mm mortars fired an illumination mission, but the men on watch still could not get a good visual identification. One of the platoon leaders reported they were ". . . picking up speed. Headed right for us."

Folsom quickly moved forward, trying to assess the situation. At last his night vision glasses revealed a small herd of camels. Folsom instructed his company "Just don't engage them."[8]

On the eastern axis 1st Marines and their attached units advanced toward as-Shatrah with Alpha/1st Tank in support of 1/4. At the small town of an-Nasr 2nd LAR became embroiled in a heavy firefight with Iraqi Army units; when Dave Banning arrived with his tank company he found a confused situation where 2nd LAR was trying to bypass the town. Wary of getting his tanks snarled in the narrow streets, he sent one tank across the long bridge over the Gharraf Canal. Discovering four enemy tanks milling about, the platoon leader called in artillery. The force dropped off infantry elements to secure various towns, and by evening 2nd LAR was at the Qalat Sukar Airfield intersection where Route 17, the main east-west axis connecting the two separated columns, branched to the west.[9]

By the morning of 27 March the 1st Marine Division was arrayed in a broad north-facing arc with two regiments along Route 1, the other on Route 7, but separated by a broad expanse of uncontrolled ground along Route 17.

The plan was now for the 1st Marines to continue their advance along Route 7 toward al-Kut. The other two regiments were positioned for a strike along Route 8 past ancient Babylon and into southern Baghdad, but that prize was reserved for the Army. Instead, the Marines would swing back east along Route 27, jump the Tigris River near an-Numaniyah, and swing west with the Tigris protecting their left flank in the final approach to Baghdad.

The diehards had not given up resisting the American advance. Aaron Klein, on Route 1: "We had seen the same vehicle several times driving past" and at a halt the vehicle approached again. "Some shots were fired. . . . We ended up getting the guys out. One of the members in there had been shot—I don't think by us—was hit in the leg. We ended up searching the vehicle, finding several Republican Guard identifications, a tracking device . . . as well as several weapons. I think we found seventeen AK-47s, RPGs, some small sidearms, a knife, bayonets. . . ."[10]

The 5th Marines were already in position along Route 27 when logistics—the true bane of military operations—halted the division yet again. The Army's V Corps, the designated main effort, had lurched to a halt from lack of adequate transport.

The Marines had pared down their logistical transport requirements by leaving most gear behind, dragging along only essential bulk commodities, reducing rations to two MREs per day, and airlifting in absolutely critical items. Dan Hughes explained that the bladder fuel "was enough that if you were getting right on the edge of your range, it'll give you that little extra to either keep you in the fight, or to extend you until you could do a deliberate refuel. . . . We never used it unless we actually needed to."

Mike Mummey, said that as S-4, "I was a cat-herder. . . . We had battalion combat trains, so we had . . . the battalion recovery vehicles, the medical team—ambulances, communicators, we had truckloads of ammunition, fuel, chow, water. . . ." Combat trains was a forward "emergency" element. "They would normally get their resupply from their Logistics Package—their Log-Pack. Their Tank Leader (typically a master Sergeant responsible for company supply) would bring that LogPack," a daily allotment of fuel, ammunition, and supplies. "Those LogPacks were generally kept back with the battalion field trains, which is where the main battalion maintenance was, BAS (Basic Allowance for Subsistence), supply; they were in a separate group farther back. They were about twenty-five clicks back. I was anywhere from six to twelve clicks back from the battalion." Resupply had been intensively practiced in training.

Delta Company/1st LAR was better prepared than most, thanks to the foresight of SSgt Jason Kappen. He had packed the company's LAV-L transports with all the MREs he could lay his hands on, and continued to accumulate more each time the logistical trains showed up. The company ended up passing out crates of meals to the infantry, who were completely out of rations.[11]

Mike Mummey also recalled no shortage of food in 1st Tank. "I hear stories of people saying, 'Oh, yeah, we were starving.' We had too much of that stuff. I think it was because their logistics people weren't out looking for it."

Problems were by no means limited to food. Vehicles were cannibalized to keep others running, and dead vehicles were towed along as rolling junkyards for spare parts. CWO Jerry Copely, the maintenance officer for Folsom's LAR company, traded a coveted pornographic magazine for spare parts. LtGen James Conway, the MEF commander, was startled to see mechanics perched precariously atop an AAV attached to the 5th Marines, working on it as it was towed along at forty-five miles an hour. He thought to himself "Now that's got to be a safety violation."[12]

Tanks too were cannibalized. Mike Mummey: "Alpha Company had a tank that went down.[13] They ended up taking it apart for parts, and then called up the Combat Logistics Battalion 18 and said 'Hey, we got a tank. You can come pick it up at such and such a grid. And good luck moving it because it's missing road wheels, it's missing track. . . .," a lengthy laundry list of parts. Even the head count "was always in flux, because you'd be evacuating something back, or had comm techs, some supply guys. . . ."

The areas to the rear were still seething with *fedayeen*. On 29 March Delta Company tanks were assigned to escort a critical supply convoy for RCT-1.[14]

To avoid revealing the plan, 5th Marines were obliged to pull back toward ad-Diwaniyah on Route 1. Much was made in the press of how the Marines were halted, and then driven back, by ferocious resistance.

Commanders inevitably worry whenever their forces are divided and vulnerable to being surprised by enemy forces lurking in the spaces between. The Marines took advantage of the halt to push in from both sides along Route 17, the only major road that connected the division's two axes of attack. From the west 3/4, supported by Captain Brian Lewis's B Company, 1st Tank advanced toward the predominantly Coptic Christian town of Afak.

The civilians greeted the Marines, pointing out the direction that *fedayeen* had fled. As the column sat in the town square and conversed with the town leaders, sporadic sniper fire came from the northern part of the town. The infantry battalion commander sent Kilo Company and a platoon of tanks to envelop the snipers from the west, while the remainder of the tanks sat on the road, ready to lay down a base of fire. The tanks surprised the insurgents, machine-gunning ten in the open. Disposal of captured enemy weapons was always a problem, one which the infantry solved by tossing them under the treads of the tanks. The infantry commander was so pleased with this tactic that he dubbed it the Afak Drill, a tactic he would utilize repeatedly in coming days.[15]

By late afternoon the task force reached the next major town, al-Budayr, whose residents also held little affection for the Saddam regime. Led by the tanks, Marines in AAVs barged into the town square, while tanks enveloped the town to cut off escape. In the square the locals urged an immediate attack on the local police station, but after blowing a hole in the wall, the Marines found that the police had fled.

On the eastern side of town the tanks on overwatch spotted four armed and uniformed men mixing with the market crowd 650 meters away. The gunner of one tank patiently waited and watched through his long-range sight until he had a clear shot, then killed one of the men with a single round from the coaxial machine gun. The crowd scattered, then returned to drag the body away. A few minutes later the remaining three soldiers reappeared, and the gunner killed a second. Some in the crowd waved at the distant tank, but the other two uniformed men never reappeared.[16]

Along the Route 1 axis *fedayeen* holdouts were still attacking units as they moved past, and D/1st LAR was ordered back to help K/3/5 clear the site of the previous night's battle. The enemy fled, but crossing the broken ground one of the dismounts, Cpl Jesus Suarez del Solar, stepped on one of the DPICM bomblets littering the ground.

SSgt Kappen quickly brought one of the LAV-Ls designated as an ambulance, dumping a load of supplies and evicting reporter Bob Woodruff's news crew to make space. As the Marines waited at the designated landing zone for the evacuation helicopter which had been delayed by mechanical failure, Woodruff asked to borrow Kappen's personal camcorder to replace one left behind where the film crew was evicted. Furious, Folsom who had never wanted Woodruff along—forbade Kappen to allow anyone to film the struggle to save del Solar's life. After a two-hour delay, it was decided to carry del Solar to the aid station by ground transport. After he watched the ambulance drive away, Folsom absent-mindedly wiped his grimy face and tasted blood—del Solar's blood. The corporal died en route.[17]

Despite the halt, the Marines did not sit in place. Patrols were sent up Route 1 and as far as practical into the marshy ground off the highways, with Charlie Company/1st Tank supporting the LAR.[18]

On 31 March the Marines resumed their advance back up Route 1 toward ad-Diwaniyah, but the Iraqi resistance had stiffened as the *fedayeen* bullied the remaining militiamen with threats to the lives of them and their families. First and 3rd LAR were leading the 1st Marine Division, seizing the airfield at Hantush. Farther back along the line of supply, 3rd Tank's security element was kept busy.

On the morning of 1 April the same infantry-heavy task force that had captured Afak and al-Budayr—K/3/4, B Company, 1st Tank, and a CAAT with Humvees—had withdrawn west and were poised on the eastern outskirts of ad-Diwaniyah. Reconnaissance suggested that Iraqi tanks were present in the city, and in the marshy ground heavy vehicles would be restricted to the hard-surface roads. The battalion commander, Lt Col Bryan McCoy, decided to use his Afak Drill, with a tank-heavy force as a base of fire and an infantry-heavy force to envelop the town.[19]

Lewis positioned his tanks atop a highway overpass about 1500 meters east of the city, the only commanding terrain in the otherwise flat alluvial plain, while most of the infantry in AAVs and a platoon of tanks continued on to a traffic circle closer to the town. The Iraqis mounted a typically disorganized defense. The tanks on the overpass began to exchange machine gun fire with positions dug into earthen berms north of the highway, in a large palm grove north of the highway, and in houses surrounding a nearby factory complex. A bus raced to and fro between the town and a traffic circle at the edge of town until the tankers spotted armed men exiting the bus and destroyed it with a main gun round. The tank optics could clearly pick out the *fedayeen*, dressed in the black clothing and ski masks that the Marines scornfully called "ninja outfits." They melted away in the face of the gunfire, and the infantry gathered in a frightened militiaman who explained that the *fedayeen* were mainly shooting militiamen who declined to fight.

The tanks on the overpass were still firing into the grove. Lewis complained that "They pop up every few minutes, like whack-a-moles. They're out there all right, only they're dug in like ticks."[20]

Kilo Company's AAVs and accompanying tanks were herringboned along the sides of the highway near the traffic circle as dismounted Marines began to search the buildings near the highway. A single recoilless gun round impacted in the road, and the vehicles scattered for cover, with the exception of the tanks. They simply sat "as though it were beneath their dignity to move," and searched for the offending gun until ordered down off the exposed road.[21]

A platoon of Marine infantry attacked the palm grove (joined by Captain Lewis and his First Sergeant on foot), while the tank platoon near the traffic circle positioned themselves to cover the ground between the grove and the town. Most of the defenders fled out the far side, only to be cut down by long-range tank fire in the open fields. The tanks then moved farther along the road, gathering up prisoners sent out by the infantry. Most were local militiamen; the *fedayeen* had as usual escaped at the first shots.

In a risky night march the division, with the 5th Marines in the van, swung northeast along Route 27 and away from Baghdad. The next major objective was to seize a crossing over the Tigris at an-Numaniyah. The 2nd Tank Battalion task force with the bulk of 5th Marines would try to seize the bridge in the city; as insurance, 2/5 led by D/1st LAR would turn left and head up the west bank along "River Road," trying to find a site for a pontoon bridge if needed. River Road branched off of Route 27 in the city, so the battalion would have to cut cross-country to intercept the road. The 7th Marines would exploit whichever force succeeded.

As the Marines probed for a crossing point, they began to encounter problems that would beleaguer them for years in the Mesopotamian valley. The 1/5 mechanized infantry team was anticipating being one of the first into Baghdad. Hayes: "We get there and apparently it wasn't a big enough bridge to support the tanks and all the vehicles. Somebody said it was just some wooden bridge. We ended up having to turn everybody around. . .

"That was probably one of the worst days in Bravo Company history."

First one tank and then another became mired, then the retriever bogged down. "That was the day that my tank went down." The tank would not start, and "It was not a very good time to fix it. After a few hours we abandoned all those vehicles. We didn't blow them up. We took all the weapons, sabotaged them to the point where they were just vehicles sitting there, took out the radios, took the firing pin out, burned all the maps, and took all our gear." Hayes ended up in a Humvee with a sniper team, and in the confusion left a prized pair of old sandals.

On the eastern axis 1st Marines finally resumed its advance north of al-Hay where Marine artillery was "pounding the living hell" out of the *Baghdad Division* at al-Kut. Banning's Alpha/1st Tanks were told to backtrack south and attach themselves to 2nd LAR for the push through al-Kut and on to Baghdad. Outside al-Kut the LAVs encountered a road obstacle that turned out to be only barbed wire, but Banning said, "My lead platoon commander called back and said he thought there might be mines up ahead because his driver was reportedly seeing hotspots in his thermals. We herringboned and the XO prepared some on-call targets in case we started taking fire. We called the engineers up and this armored combat earthmover (ACE) drives right through our lines. We had told them that my tank was the one with the big six spraypainted on the panel on the back of the turret and that he should stop there. But he drove right through our lines and right into this thing. Sure enough, he hits a mine and blows his track off. I couldn't believe that that just happened. The guy starts jockeying around in his ACE and my platoon commander is telling me on the radio to tell him to stop because he was chewing up more mines. Sure enough, you look up there and as his tracks were moving there were more mines kicking up. It was a miracle the guy wasn't killed. So we get him to stop and get him out of there. We bring the rest of the engineers up to blow these mines in place. It took 40 minutes; it was excruciatingly slow."[22]

On 2 April the main force of 2nd Tank "bounced" the Tigris bridge, seizing it with only minor damage by RPGs and destroying several tanks and BMPs in the streets. The 3/5 held a narrow passage through the city to allow the tanks and LAVs of 1st LAR passage and maintain the momentum of the attack. The 7th Marines quickly closed up to secure the city while 5th Marines pressed onward with 2nd Tank in the lead.

By 3 April the main body of the 5th Marines had turned west toward

Baghdad. The main push was to be along the highway north of the river. The 1st Marines, along with two battalions of RCT-7 (1/7 and 3/4) surrounded the *Baghdad Division* at al-Kut from the south and north, respectively. Al-Kut sat astride Route 7, potentially the main supply and communications axis for the division, between an-Nasiriyah and Saba. In the interim units were supplied with fuel brought into the captured airfield at an-Numaniyah by C-130 transports.

Dave Banning's company was spread up and down the eastern highway, with only one platoon and the headquarters section south of al-Kut. "I start to get a little nervous because the town was getting more built up and we had no attached infantry. I had the attached LAV platoon but still no dismounted infantry. I radioed back to tell them we were headed into a built-up area and that I needed some infantry ready to do something." He knew that 7th Marines were to attack from west of the town, but could not contact B Company tanks. Senior officers were conferring nearby, so Banning walked over to eavesdrop when "All of a sudden, this infantry battalion blows right through us and up towards this urban area. In Al Kut there is this big military compound in the south, which is kind of open, but then you can see the typical Iraqi townhouses and all that stuff. Things got more built up towards the north. These infantry guys started heading pedal to the metal into this urban area. I hopped back on my tank and told my guys to stand by. Sure enough, as soon as they turned the corner there was a hail of gunfire. I punched my first two tanks out there to go reinforce them but I didn't want to leave the intersection unsecured." When an LAR company arrived to secure the intersection, Banning headed his small tank force into the town.

"My lead platoon commander was whacking these Iraqi reinforcements coming over this bridge. They were just streaming over this bridge like they were completely oblivious to the fact that there were two tanks sitting at the bottom of it, just killing them like cancer. This went on for 10 or 15 minutes and then everything quieted down. There was this lull in the battle and we figured we'd drive over the bridge, go link up and head out, but then we get the word to withdraw to the south. So we withdraw back south, get the order that now we're going to march down around Numaniyah and we're going to link up in Numaniyah with the rest of the division."

Captain Brian Lewis's B/1st Tank was ordered not to enter al-Kut, but provided long-range fire support. However, Lewis was shot in the hand in another of the random ambushes outside the city—what he called "a cheap lesson in how the enemy was operating." [23]

The delay in clearing al-Kut and bringing RCT-1 across the Tigris caused the division commander, Major General John Mattis, to conclude that Colonel Joe Dowdy, the CO of the regiment, was not being sufficiently aggressive. Mattis literally "kicked him upstairs"; Dowdy was made senior officer airborne to coordinate activities from a specially equipped P-3 Orion, replaced by Colonel John Toolan.[24]

Parts of the Republican Guard *al-Nidah Division* chose to make a stand at the town of al-Aziziyah, astride Route 6 about 60 kilometers northwest of the junction with Routes 7 and 27 at Sabat. Tanks from A Company, 2nd Tank Battalion under Captain Todd Sudmeyer (a former enlisted Marine) shot their way through enemy fire and along the main highway that bypassed the town proper, but the long columns of AAVs and soft-skinned vehicles behind could not brave the fire, inaccurate as it was, coming from the town.

Long-range RPG fire also came from behind a taxi parked near the canal south of town, skipping across the road near the halted AAVs. Major Andrew Bianca, the XO of 2nd Tank, had his gunner fire on the taxi. The high-velocity round tore through the taxi with no apparent effect, but the fire ceased and a few moments later the car burst into flame.[25] Meanwhile Sudmeyer's tanks continued forward, quickly overrunning an Iraqi military post after a brief firefight.

Now 3/5 had to dismount and clear the town, supported by mortars, artillery, AAVs, and air strikes. As in so many incidents, the residents told the Marines that the Iraqi soldiers fled at the sight of Marines crossing the bridge that led into the town. The infantry went door-to-door, clearing the city. The 1st LAR screened the exits from the city, trying to intercept fleeing enemy. Warrant Officer Mike Musselman, the crew chief for Lt. Col. Duffy White's vehicle, spotted several Iraqis shucking off their uniforms about 1500 meters from the road. When one ill-advisedly fired at the LAV, the return 25mm cannon fire made him turn "kind of misty." The others surrendered to the platoon's dismounts.[26]

The tiny town proved to be fateful for 2nd Tank. Daniel Benz:

". . . our scout platoon commander got wounded and a bunch of scouts got wounded. We took our TOW platoon commander and put him in change of the scouts. I had trained him on all of the scout platoon tasks. They have the scouts out front, which in retrospect is, "What the fuck were they thinking?" A scout platoon is normally like, let's seek opportunities, but it's already a predetermined course of action. And what can the scouts see that a tank can't see? The scout platoon commander gets shot in the head. KIA instantly . . . first day on the job. Then we had three other KIAs that day in the battalion and really a lot of WIAs. One of our company commanders, his tank got shot and was on fire and the ammo cooked off. He bailed out of the tank no problem. But he gets halfway across the street, he's running for some other tank, realizes he forgot his map, so he runs back across the street, climbs up in the tank, gets his map, gets down, gets shot right—it was one of those weird things: ricocheted off his shoulder blade, went up through his mouth, came out, bounced off a board, down through his throat. Whatever. He lived. He was okay. He went home. It was probably the worst day for the battalion."[27]

South of the river the Marines encountered resistance from *fedayeen* fleeing the fight in al-Aziziyah, but the River Road mission turned out to be misguided. The long column struggled in a maze of single-lane roads. A yellow pickup truck turned to run from the LAVs, but a tarp over the cargo bed blew off revealing a mounted heavy weapon. Two LAVs opened fire, and Lt Brandon Schwartz recalled the look of fear on the driver's face, and seeing him blown into the passenger's lap by the impact of the 25mm rounds. The truck burst into flame and careened into a drainage canal parallel to the road, where stored ammunition began to cook off.[28]

The column, led by 3rd Platoon of the LAR company approached a nameless village, simply a place where the road made a bend away from the river. Most of the crews had been lulled by heat, tedium, and fatigue when suddenly small arms fire and RPGs began to impact all around. The platoon was too closely engaged in the L-shaped ambush for air or artillery support, and backed down the roadway, hammering away at trees and buildings that sheltered the attackers. Two more platoons of LAVs came forward, and the 81mm mortars began to drop rounds on the retreating *fedayeen*. Within minutes artillery, Cobra gunships, and F-18s arrived to add to the destruction.

Inexplicably an Iraqi family in a sedan—a white rag tied to an outside mirror—drove the length of the column of LAVs and continued on into the ambush site before seeing the volcano of dirt and debris rising from the village. Pulling a quick U-turn, they raced back down the firing line and out of the battle. While assessing the fight, Folsom received orders to backtrack and cross the river at an-Numaniyah that night.[29]

The path to Baghdad was a four-lane divided highway, and the 5th RCT with 2nd Tank Battalion, took the lead. First Sergeant Edward "Horsehead" Smith of F/2/5 told Bing West that ". . . we're going by so fast that the tanks are shooting only at enemy main systems, like other tanks." He went on to explain that it was hard to keep up his men's morale, since the tanks were taking all the shots and the infantry was cooped up in vehicles.[30] The main problem, as usual, was supply since the logistical units were fighting a different kind of war against distance and traffic jams.

At 1100 on the morning of 4 April the double column moved out, with B Company 2nd Tank on the left side of the highway median, followed by A Company, a column of 34 tanks. To the right an equally powerful column of CAAT Humvees advanced in parallel. Remnants of the *al-Nidah Division* were still entrenched in bunkers dug into berms along both sides of the road, and the column was under steady fire from small arms and the occasional anti-tank missile.[31]

Most of the fighting was with machine guns and rifles, though occasionally the tanks would fire their main guns at particularly troublesome positions. A few tanks and BMPs were destroyed sitting in place or fleeing the area, but the main obstacles were burning vehicles hit by aircraft.

About 35kilometers up the highway, near Tuwayhah, the highway narrowed and the tanks moved on in a single column, with Charlie Company taking the lead. *Al-Nidah* might have collapsed, but the *fedayeen* were still full of fight. Old vehicles were stacked as barriers, and an oil-filled trench was set afire: it was more a psychological then physical barrier, but 2nd Lieutenant Adam Markley still found it "scary as shit. I didn't know what was on the other side."[32]

With so many rounds flying through the air, the Iraqis were bound to score a few lucky hits. The heaviest fire was falling on company XO 1st Lt Charles Nicol Jr's tank. ""He's got five or six antenna on his tank—it's like having a giant 'shoot me here' sign," Lt. Markley said.[33]

One hit ruptured a fuel bladder hanging from the turret rack of the tank belonging to Captain Jeffrey Houston, the CO of A Company. It started a minor fire, and fuel flooding into the air intakes stalled the engine. Houston dismounted and ran to another tank, where he could use the radios to control his company. The driver, LCpl Billy W. Peixotto, dismounted to activate the external fire extinguisher. More enemy fire poured down on the tanks, Humvees, and AAVs. A rifleman was seriously wounded, and the commander of the Scout Platoon, Lt. Brian McPhillips, was killed by a round to the head.[34]

Most people think that tankers are always protected by their thick armor, but since the earliest days tankers have realized that their best protection is awareness of the enemy's position. Usually the tank commander fights standing in his open hatch, and in a running gun battle like that along Route 6 the loaders were using their roof-mounted machine guns to protect their tanks. A few kilometers farther along a random RPG hit the open loader's hatch of Markley's tank. The blast deflected downward, mortally wounded Cpl Bernard G. Gooden, and disabled the tank's radios. The blast disabled the turret traverse, stunned Markley, and knocked his portable GPS unit overboard. In the ensuing confusion he missed a turn, and headed into a maze of narrow roads. The rest of the long column followed.

The Marines quickly realized the problem, but the 15-kilometer long column was shortening like an accordion as more vehicles pushed into the jam. Markley had another tank commander plot their location; when gunner Corporal Julio Cesare Martinez restored the radios, he notified Nicol of the mistaken turn. The unwieldy column would have to be turned around on itself in the narrow streets, under heavy fire. Martinez went back to work, traversing the heavy turret with the hand crank, and firing the coaxial machine gun with the manual trigger.

Captain Sudmeyer veered out of the column and raced to the front, turning his tank broadside in the road to protect other vehicles as they made the risky U-turn. An infantryman in one of the AAVs was hit. Captain David Bardorf raced up with an ambulance to evacuate the casualty, adding more unavoidable confusion.[35]

Back on Route 6 Peixotto was trying to restart the tank engine, but it still would not start, so he climbed out again into the small arms fire and RPGs sweeping the road. Gunner Cpl Alfredo Ramirez and the loader manned the turret roof machine guns and returned fire. The loader's machine gun jammed, so Cpl Michael Ackerman used a rifle and pistol to hold the *fedayeen* at bay.

Houston ran back to his tank and got on the tank-infantry phone attached to its rear. A random round struck him in the jaw and he collapsed in the road, bleeding profusely from a severed artery. Peixotto raced to his side. "Because of the way you train as a Marine, you don't think about yourself. You basically, you think about others. You see somebody down like that, I mean no matter who you are, you think about yourself for a brief second, and you decide, you go do it."[36]

Ramirez and Ackerman dismounted to help Peixotto and cut loose the troublesome fuel bladder. They tossed Peixotto all the extra clips of pistol ammo, and first aid compresses.

Encouraged by their modest success, *fedayeen* and soldiers hiding in the ditches and berms rushed the tank, closing to within 20 meters.

Peixotto dragged Houston up against the side of the disabled tank, where "It felt like being inside a building during a thunderstorm. It's scary, but you really don't think about it. There was another Marine down, and I didn't want to be the one who could have helped him and didn't." Holding a battle dressing to Houston's shattered face with one hand and firing his 9mm pistol with the other, Peixotto talked over a radio held to his face by another Marine, coordinating the protection of his tank and captain.[37] He emptied clip after clip of ammo at the enemy.

Colonel Joe Dunford's regimental command group arrived at the scene, adding their modest firepower and positioning vehicles to help shield Houston and Peixotto. The ubiquitous Bardorf arrived with the ambulance and battalion surgeon Navy Lt. Bruce Webb. As the crew helped Houston—barely conscious—into the ambulance, the driver, Cpl Luke Holden, was shot through the hand. Navy corpsman Thomas Smith somehow drove, held a compress over Holden's hand, and fired an M-16 out the window as they raced back through the ambush. The ambulance drove a short distance north out of the main firefight to a position where a medical evacuation helicopter could safely land. But safety was a relative thing; Holden was shot through the other hand at the landing zone.

By this time Cobra gunships were strafing and rocketing the enemy positions in the fields on either side of the road. Dunford ordered 3/5 to dismount and clear the fields, sending India Company to the south side, Lima to the north. The infantry battle among the ditches and berms consumed the next three hours, but by 1600hours the remnants of the *al-Nidah Division* and *fedayeen* had been eliminated. Dunford ordered Peixotto, Ramirez, and Ackerman to abandon the still-smoldering tank and leave with his command group.

The safety features built into the M1 tanks were glaringly obvious in this incident. Dan Hughes, still attached to 3/5, ". . . didn't notice any explosions from the tank. It basically just melted down." Hughes was told to drop the bladders from his tanks.

Despite the best efforts of the embedded reporters, the consequences of the media's insatiable desire for "breaking news" could be agonizing for family members at home. Peixotto:

> When it happened, I wrote my dad a letter telling him not to tell my mom or anything. They had told me not to do anything stupid, and I did.
>
> Before they got the letter they found out about it. I actually had been pronounced killed in action over the radio. I think the whole crew had been. Nobody knew where we were. My mom called, and all a woman would say was that they had no information. It was about three or four days of about pure hell for my parents.
>
> If I could do it over again, I'd tell him [Houston] to stay in the tank.[38]

Jim Landers of the *Dallas News* reported:

> "There were enough rifles, RPGs, and other small arms in that town to outfit an entire Marine division—15 buildings' worth," said Lt. Col. Mike Oehl, 2nd Tank Battalion's commanding officer. "These were Islamic Jihad guys from all over the Arab world. We have intelligence reports that they've been staying at the Sheraton in downtown Baghdad."[39]

The next day, now far to the division rear, Folsom's LAR company struggled through the usual traffic jam, and was eventually assigned to investigate and clear weapons depots and military facilities. Engineers were destroying massive quantities of weapons, and the LAVs disabled parked armored vehicles with cannon fire, but there was just too much ammunition to be safely or practically destroyed.

Securing the "rear areas" was a foretaste of things to come in a country where everyone seemed to own an AK-47. Mike Mummey said, "When the tanks would move through, it was like parting the Red Sea. They would all fall back, but then as soon as the tanks left they would . . . start shooting again." The security platoon did little fighting, but "That was the first time I saw a TOW missile being used for an anti-personnel weapon. One of our TOW guys, there was a guy shooting at us, and he just decided to go ahead and let a missile fly. It stopped him from shooting."

Securing the rear provided a strange cultural shock to Folsom's men. His 1st Platoon was positioned near a factory where a dog was caught and crushed between an iron gate and a concrete wall. Finally the Marines decided to bury the stinking, maggot ridden carcass. The local idlers would only look on in amazement and contempt for those who would stoop to bury an unclean animal.[40]

By now the unrelenting action had begun to take a toll on machines as well. Banning's company was towing two disabled tanks and an M88 on a 250km road march. "By the time we limped into an-Numaniyah, we were hurting. I reported in when we got there and the regimental commander looked at his watch and told me we were crossing the line of departure (LD) for the attack in 20 minutes. I asked if he really needed us. At this point, two other regimental combat teams had already passed through the area. I said if he needed us we'd go but, by the time we're done, he would only have three functional tanks. I had tried everything, every communication pipe available, to get a replacement engine for that tank. It was very frustrating because we saw a helicopter land next to one of the infantry battalions and somebody hops off, delivers an ammo can to one of the Marines and we can see them delivering .50 caliber parts—with a helicopter to this infantry battalion. I was sitting there pulling my hair out. I was towing a tank! This was one-thirteenth of their major combat power and they couldn't get me an engine for this thing? I thought that was crazy. We linked up with the support guys and explained the situation. The lieutenant colonel in charge points to this row of CONEX boxes on the back of trucks and told us the parts were in there somewhere. Super. What a big help. We took the tank that had the worst suspension and roped it to the ground. We spent 24 hours playing parts swap games and got the rest of the tanks in the company fully mission capable. By the time we left in the morning, we had 12 tanks full up and ready to go."[41]

NOTES

1 As quoted in Landers, *The Marines' 2nd Tank Battalion Used Speed and Armor to Make Quick Work of Saddam Hussein's Regime.*

2 In the aftermath of this incident several conflicting accounts were published. This account is drawn from the most reliable, Anonymous, Operation Iraqi Freedom Iraq—1st Tank Battalion, After Action Report.

3 Livingston, *An Nasiriyah*, p. 176–177.

4 Folsom, *The Highway War*, p. 145–146.

5 Landers, *The Marines' 2nd Tank Battalion Used Speed and Armor to Make Quick Work of Saddam Hussein's Regime.*

6 Galuzska, *Interview with Major Daniel Benz*, p. 24.

7 Harris, *Interview with Second Lieutenant Aaron Klein.*

8 Folsom, *The Highway War*, p.158–159.

9 Lessard, *Interview with Major Dave Banning*, p. 12.

10 Harris, *Interview with Second Lieutenant Aaron Klein.*

11 Folsom, *The Highway War*, p.163.

12 Ibid, p.163–164.

13 A-41, the tank hit by friendly fire.

14 Anonymous, *Operation Iraqi Freedom Iraq—1st Tank Battalion, After Action Report.* p.25.

15 West and Smith, *The March Up*, p. 90–93.

16 Ibid, p.95–96.

17 Folsom, *The Highway War*, p.170–179.

18 Anonymous, *Operation Iraqi Freedom Iraq—1st Tank Battalion, After Action Report.* p.26–27.

19 The following account of the an-Diwaniyah action is drawn primarily from West and Smith, *The March Up*, p. 113–123.

20 West and Smith, *The March Up*, p. 118.

21 Ibid, p.117–118.

22 Lessard, *Interview with Major Dave Banning*, p. 13.

23 Anonymous, *Operation Iraqi Freedom Iraq—1st Tank Battalion, After Action Report.* p.29.

24 West and Smith, *The March Up*, p. 149.

25 Landers, *The Marines' 2nd Tank Battalion Used Speed and Armor to Make Quick Work of Saddam Hussein's Regime.*

26 West and Smith, *The March Up*, p. 151, 172.

27 Galuzska, *Interview with Major Daniel Benz*, p. 28.

28 Folsom, *The Highway War*, p. 224.

29 Ibid, p.230 241.

30 West and Smith, *The March Up*, p. 154–155. Horsehead Smith was killed in action on 3 April.

31 Ibid, p. 151, 157.

32 The following general description of the action and quotations, unless otherwise noted, are taken from Landers, *Ambush Costly for Marine Battalion*.

33 Landers, *Ambush Costly for Marine Battalion*.

34 West and Smith, *The March Up*, p.159–160.

35 The rifleman later died of his wound. Ibid, p.160–161.

36 Youngquist, *Pulled Through by His Buddies*.

37 Quoted in Steinkopff, *Five Marines Honored For Service in Iraq War*. Peixotto was eventually awarded the Bronze Star.

38 Youngquist, *Pulled Through by His Buddies*, Winston-Salem Journal, September 26, 2009.

39 Landers, *Ambush Costly for Marine Battalion*.

40 In Islam the dog, like the pig, is unclean because of its habits of eating carrion and excrement. Folsom, *The Highway War*, p. 251.

41 Lessard, *Interview with Major Dave Banning*, p. 13.

CHAPTER EIGHT

The Prize

Hell, these are Marines. Men like them held
Guadalcanal and took Iwo Jima. Baghdad ain't
shit.—Major General John F. Kelly, USMC

N OW ONLY THE NAHR Diyala River, with a deep channel and high, steep banks, stood between the Marines and Baghdad. With the division cut loose from any continuous logistical train, Houston's crippled tank was destroyed in place.

The remnants of the *al-Nidah* were a less effective obstacle. Shortly after the 5th Marines went into defensive circle for the night three 122mm rockets exploded harmlessly within the position. They were answered by six dozen rounds from the 11th Marines' 155mm guns, dumping nearly 8000 DPICM bomblets on the firing site detected by counter-battery radar. There were no more rockets.[1]

After two and one-half hours of sleep Folsom's D/1st LAR, joined by A/4th LAR who had finally received their vehicles and caught up with the division, was ordered to join the rest of the battalion at the front of the division. The LAVs inched their way through the massive jam along Route 6; vehicles were parked haphazardly and exhausted men were sleeping everywhere, even on the pavement.[2]

No enemy armored threat had materialized, and for the assault on the city the tanks were broken down even further. "Tiger" (1st Tank) would give up one of its two remaining line companies, Charlie, to be split between operational control of 1/7 and 3/7. In return an additional infantry company, A/1/7, would pass to control of the tank battalion to create an infantry-heavy mechanized task force, Task Force MECH.[3]

On the morning of 5 April 3/4, supported by B/1st Tank, attacked straight into the built up area on the east shore of the river, advancing up to the four-lane highway bridge. While higher commands were debating the capture of the bridge and how it fit into the larger plan, the Iraqis blew the center span. The demolition engineers were inept, and although it was impassable to vehicles, infantry could force a crossing.

LtCol Duffy White's 1st LAR had already been unleashed to perform one of its primary tasks, to search for a suitable crossing point over the river if the Iraqis blew the main bridge. Each company was assigned a sector of the river bank, and Dragon Eyes were launched to speed the search.

As the Marines waited, Bing West and Ray Smith walked over to where a crowd had gathered. A T-72, fleeing the previous afternoon's fighting, had forced a bus into a drainage canal, then itself overturned into the canal, drowning the crew. The Marines called up an M88 to extricate the bus, to the delight of the owner-driver.[4]

Selection of a bridging site was delayed as the LAR companies struggled in the maze of animal paths and poor roads in the river plain, and engaged Iraqi units on the far side of the river. The frustrated reconnaissance Marines were no more pleased when two Navy F-14s made a visual pass over the Weapons Company, then dropped two large bombs which fortunately were wide of their target.[5]

By late afternoon three battalions were in the eastern outskirts of Baghdad. Task Force MECH, led by L/3/4, would cross at the southern bridge. The supporting tanks provided covering fires, and destroyed one BMP-2.[6]

Bridging assets were sent five kilometers north of the main bridge to a site selected by Task Force BLADE (with Dave Banning's A/1st Tank and 3/7), in case it proved necessary to repair a destroyed bridge. The plan was for the highly-visible tanks to fix the enemy's attention while the infantry forced a crossing elsewhere to secure a bridge site.

On the morning of 6 April K/3/4, supported by a platoon of C Company tanks, made final preparations to rush a pedestrian bridge near the cratered span. As the 155mm howitzers fired a "danger close" fire support mission the Iraqis returned artillery fire, leading to confusion when the Marines thought their own artillery was falling short. The infantry rushed the bridge and quickly forced a crossing, throwing a large metal gate and planks across a gap blown in the bridge. Within minutes India and Kilo Companies were clearing the warren of houses that nearly overhung the steep bank.[7] Engineers repaired the bridge at the alternate BLADE site, a pontoon bridge capable of carrying heavy vehicles was quickly built near the damaged highway bridge, and the highway bridge itself was under repair. The Marines now had three bridge sites.

Engineers, creeping through a drainage tunnel to a position under the bridge, determined that the highway bridge at Task Force MECH's crossing site could be used. Inept enemy demolitions had cratered the decking, but the supporting steel stringers were intact. On the morning of 7 April tanks that had been suppressing fire from the far shore backed off to make way for an AVLB. The artillery smoke fire mission, intended to obscure enemy observation, was abruptly called off, but mortars took up the slack. The AVLB moved onto the bridge, and quickly emplaced its span.

Though this crossing was intended as a feint, LtCol Jim Chartier demanded—and received—permission to make this crossing site the main push toward the city.[8]

By 1230hours Delta Company tanks had encountered and destroyed two T-72s, but enemy resistance was collapsing. The main hindrance was the ubiquitous mines. The remainder of the day was spent securing objectives in the outskirts of the city, and trying to deal with the flood of refugees and heavily-burdened looters flowing around and through Marine positions.

With major Coalition forces now arrayed around the city, it was no longer an option to ignore disagreement over how to control a city the size of Baghdad. The US Army, the first to arrive, had already commenced its vaunted "thunder runs" into the city. Heavily mechanized but short on infantry dismounts, the concept was to send powerful armored columns into the city, shoot up any resistance that revealed itself, and then withdraw, repeating the process over and over. The idea was simply to inflict casualties with the hope of wearing down any resistance in the city. The problem of course was that the city was full of innocent civilians, many of whom would inevitably become casualties. Their friends and relatives would then become new enemies, fueling the very guerrilla movement the Coalition hoped to avoid.

At the other end of the spectrum were the British. Building on experience in Northern Ireland, their model was more akin to police than military action. They wanted to encircle the city and carefully—if slowly—build relationships with local residents, identify enemy leaders (many of whom were at this point either Ba'athist die-hards or foreigners not much beloved by the locals), and decapitate the resistance with selective raids.

In the middle were the Marines, who also wanted to build upon prior experience. Their plan was a variant of the "spreading ink blot" approach, which in some circles was held in ill repute after being prematurely aborted by William Westmoreland in Vietnam. The idea was to forcefully seize and hold critical positions inside the city, destroy insurgents who would inevitably flow toward these centers, and then gradually extend control over the remaining parts of the city. The first step would be to advance along the main streets from southeast to northwest, throwing a loose net over the central part of the city.

The problem for the Marines was that neither electronic intercepts (ELINT) nor human intelligence (HUMINT) could identify any real centers of resistance. There were no centers of organized resistance. The regime had suffered a catastrophic collapse, and the *fedayeen* were inherently disorganized.

While the planners deliberated on 8 April, the Marines paused to tighten their ring around eastern Baghdad.

Like most cities, Baghdad presented a spectrum of faces. At one end were the affluent areas of office buildings, walled residences, and palatial homes, all set along broad boulevards. At the others were warrens of tiny streets, ancient or modern slums, some without basic utilities such as clean water and sewers. Both sheltered enemy fighters.

When they moved into the city on 9 April, "We had probed the southern edge of Baghdad the day before with the TOW platoon and it seemed as if the civilians were more or less disinterested, which we hoped would preclude at the least the unrestricted use of civilian dwellings for ambush sites. Of course, this turned out to be the case exactly. You really can't fight a guerilla war if the locals don't like you, and we found out pretty quickly during our attack that most Iraqi citizens thought 'Saddam bad! Bush good!'" (Maj Mike Purcell, 1st Tank Battalion Operations Officer).[9]

First Tank's Task Force Mech found itself near a large amusement park, and the Martyr's Monument to the half-million casualties of Sadaam's war with Iran. Critics would later charge that the Marines looted the monument, but reporter Elliott Blair Smith said, "As the first American inside, I can say it already was trashed." Looters had even brought their own transport. When LtCol Chartier fired his pistol into the air in a vain attempt to dissuade looters, nervous grunts whirled and fired on him. "I get peppered around my head. A paint chip goes off my nose."[10]

The infantry divvied the city up into battalion sectors, each with a few supporting tanks. Mike Mummey: "I'd end up running parts or personnel to different units. . . . Take a fuel truck over there to (re)fuel them. Once we got to Baghdad they chopped up the tanks. The grunts needed tanks, so hell, there's headquarters sections—a CO and an XO of a company—working separate from their company. You got a lance-corporal telling them where to go." Mummey's job was now to send his vehicles, usually unescorted, to assist the small units scattered about the city."

He quickly found that "Grunts aren't very good at logistics." When tanks were attached to infantry "Pretty soon, 'Hey, Top. We need fuel for these amtracs,' because they weren't getting support." With even the headquarters elements of the AAV units tied up in combat missions, support for the tanks and amtracs fell to the infantry battalions, who were not equipped to handle the added load. "You say you need fuel, and they'll bring you a five-gallon can. What am I supposed to do, light the tank on fire with this?"

The logistics chief for an infantry battalion: ". . . all of a sudden he's got a company of amtracs, then all of a sudden he's got a platoon of tanks, and pretty soon they're overwhelmed."

Advancing along the main streets and securing the dominating buildings meant days of slow, hot, boring and occasionally terrifying work as vehicles stood guard in the streets and infantry worked their way house by house, scaling garden walls, smashing through gates and doors, climbing stairs, searching rooms. Neither the Iraqi Army nor the *fedayeen* had organized themselves in any way to defend the city, but any random close-range encounter might be with either innocent civilians or the occasional fanatic who wanted to kill you.

West and Smith witnessed one such inexplicable event as a nondescript car approached a tank sited to cover a long boulevard, and the gunner fired several warning shots. Rather than steer away the car accelerated directly at the tank as Marines scrambled for cover, assuming it was suicide car bomber. The tank and a nearby AAV raked the car with machine gun fire, and everyone cringed in anticipation of the explosion.

Engineers searching the car found only two dead middle-aged men in denim work clothing. Had the driver simply panicked at the sight of the tank or the gunfire? It was just one of far too many such incidents.[11]

But there were fighters who made even the most routine activities fraught with danger. One tank commander dismounted and walked into a vacant lot to answer an urgent call of nature, only to come face to face with three armed men who popped up from among the litter. He bolted—or rather waddled at high speed with his trousers down—back toward the tanks shouting "Shoot! Shoot!" The startled loader emptied a machine gun and then his pistol killing one of the men. Nearby infantry eventually hunted down and killed the others.[12]

A prize awarded to 1/7 was to search a peninsula formed by a bend of the Tigris, which held several of Saddam's palaces, the embassy district, upper class homes, and Baghdad University at the tip. The shape resembled a cartoon scrawled on a men's room wall, so the staff planners discreetly dubbed the peninsula The Snoozle. CIA intelligence suggested that the University was occupied by large numbers of *fedayeen*.[13]

The infantry-heavy task force—1/7, led by C Company tanks—moved down the broad streets through the embassy district, peeling off platoons to secure key intersections and objectives. The tanks motored through to secure the bridge at the tip of the peninsula, and took up firing positions on the bridge ramps, from where they could dominate the nearby University campus and block any escape routes. Infantry cleared the campus, which harbored only a few die-hard *fedayeen*.

Other battalions pushed northwest and deeper into the city. Firdos Square, dominated by a huge metal statue of Saddam Hussein, lay in the path of 3/4. The fall of the statue, pulled down by an M88 from 1st Tank in front of news cameras and cheering Iraqis, quickly came to symbolize the fall of the regime to most of the world. Patrols fanned out through the city to search palaces belonging to leaders of the fallen regime, where necessary using tanks to knock down the heavy iron gates that barred entrances to palace compounds.

The CIA believed it had credible intelligence that Saddam himself was in the Almilyah Palace, on the great bend of the river north of The Snoozle, so 1/5 was assigned to storm the huge compound, actually six palaces and sprawling gardens. A platoon from 2nd Tank was first in, smashing down the massive gates, and the Alpha Company infantry spread out to search the grounds. In the close quarters one tank suffered an RPG hit in the transmission cooler—the one vulnerable spot—and quickly ground to a stop.

Once again, the modular construction of the tank proved its worth. Benz: "We had another tank from Alpha Company went into an irrigation canal. And one of my best moments was seeing my maintenance guys—they pulled both engines and sat them on a highway. Within four hours from the time when they pulled those tanks, they put the wet transmission, which still works, in the blown up tank. That one was toast. But within four hours, they had a combat capable tank back in the fight. That was pretty cool."[14]

The entire district proved to be crawling with *fedayeen*. The tanks and infantry fought their way toward the Hanifah Mosque, the tanks blasting buildings with 120mm cannon fire and the riflemen rooting out any survivors who did not flee.

Back on the eastern bank of the Nahr Diyala, with no real role in the street fighting, 1st LAR stewed and discipline suffered. That night Folsom spotted an orange flame in the back of one of his mortar vehicles, and found the crew cooking on a gas stove amid the stored ammunition. No sooner had he finished lighting into the platoon sergeant when the sound from a SAW ripped through the position. On another mortar carrier an accidental discharge wrecked the engine and sent shrapnel into one of the crew. The exasperated Folsom relieved the platoon leader.[15]

First LAR finally inched across the undulating ribbon bridge, one vehicle at a time, and was assigned to send a patrol north of the city, with strict orders to return before darkness.

By 10 April Army General Eric Shinseki's assertion that more troops would be needed to control the country than to capture it was validated: infantry battalions were being sucked into the vacuum of Baghdad.[16] First LAR had to assume partial responsibility for isolating the sprawling capital. The battalion was assigned a broad arc north of the city, between the Tigris and the Diyalah and bordering the vast slum known as Saddam City (later to be known as Sadr City). The area was littered with abandoned weapons from small arms to artillery and tanks, with dangling electrical lines thrown in for good measure. The tank battalion's M88s dragged away numerous armored vehicles.[17]

More Marine units moved in to occupy the city. Bill Hayes was now just a passenger in a heavy cargo truck. "We drove to Baghdad and met up at this big university there." Driving into Baghdad, the Marines were greeted as liberators. "These people were ecstatic. You could tell. You can't fake that emotion." Nevertheless, leaders like Dan Hughes were warning the Marines, "Don't get too friendly with them. . . . Don't be hostile, but keep them out of our area, keep them a good safety distance away, don't let them crawl on your vehicles."

Hayes thought the whole affair "Kind of anti-climactic. I kind of engaged the enemy, I guess. Good enough." Hayes could not have anticipated what was to follow. "I never in a million years thought we'd have gone back, that [it] would have lasted I guess eight more years." Hayes and others sat at the University, then at a new base nearby, while the policy makers tried— like the dog who chased cars—to determine what to do with Iraq now that they had it.

As the Marines transitioned to security duties, they found that locals had often taken neighborhood security into their own hands, and welcomed the Marines' efforts. In a peculiarity of the rules of engagement, Marines were allowed to fire upon such friendly "neighborhood watch" groups, but not at unarmed looters. Efforts to get such necessary functions as government ministries running again were equally chaotic. (1st Tank's Field Trains were responsible for the important Ministry of Oil building.)[18]

On 12 April the 1st Marine Division organized Task Force TRIPOLI to secure Saddam Hussein's reputed birthplace, Tikrit. This expedition would extend the division's area of responsibility significantly inland. The best route for the advance would require passing across the rear of two Army divisions, thus delays in massive traffic jams. The alternative, from eastern Baghdad would require the task force to negotiate a maze of small roads scouted by SEAL teams, and numerous bridges of dubious capacity. The mechanized task force therefore relied entirely upon wheeled vehicles. Built around an entire regiment of LAVs (1st, 2nd, and 3rd LAR), the force included G Company and the CAAT from 3/23, artillery, and the division Jump HQ.

On 13 April the new Task Force received credible intelligence that American prisoners were being held in Samarra, and a raid was hastily organized. After a brief battle on the outskirts the LAR battalions surrounded the town and infantry recovered seven POWs—two helicopter pilots and five survivors of the 507th Maintenance Company captured in an-Nasiriyah.

A local policeman used a portable GPS to mark the house where the prisoners were held, and the prisoners themselves persuaded their guards to flee at the approach of the Marines.

Third LAR was left to secure the town while the rest pushed on. Scattered bands of *fedayeen* staged numerous small ambushes. In one encounter Folsom watched as a large group of civilians and a few soldiers caught between the Marines and *fedayeen* milled about as if uncertain what to do. One soldier came toward the Marines, hands in the air. Gunfire scattered the civilians, but the lone soldier took a round or bit of road surface thrown up by the firing, and collapsed spewing blood from a head wound. The dismounts grabbed him and dragged him to the medical evacuation point as the LAVs and dismounts swept past.

A military truck screeched to a stop and disgorged men clad in civilian clothes who made ready to fire on the Marines, only to be cut down by sheets of fire from the LAVs. Three survivors fled into a culvert under the highway. Dismounts surrounded the mouths of the culvert, tossing three grenades inside. Folsom arrived and ordered the scouts to toss more grenades inside. Still the Iraqis held out.

Folsom dismounted and took over the M240 turret roof machine gun that had been passed to the men on the ground, but when they tried to make a concerted rush, the weapon misfired. Two more rushes fell back as the machine gun repeatedly malfunctioned. After the third rush the *fedayeen* inside replied with a grenade, wounding one scout. Someone in the tunnel shouted "Go away" and "I throw bomb" repeatedly.

The Marines could see into the culvert, where two *fedayeen* were down, badly wounded or dead, and a third was slumped against the wall holding a grenade. The man refused to surrender. Folsom shot him with his pistol, and he collapsed against the floor. After making sure the three men were dead, the Marines withdrew back up to the road, where the LAR and infantry battalion commanders were trying to establish what was going on.

Folsom was later confronted by his First Sergeant, Ruben Guzman, who told him, "You gotta stop this shit, sir. You're gonna get yourself killed." Guzman, better than Folsom, realized the implications of having the CO killed or wounded.[19]

In Tikrit he Marines fanned out to secure various objectives with far less resistance than expected. Delta Company, 1st LAR drew a plum prize, one of Saddam's palaces perched on a bluff overlooking the Tigris. Ordered to secure the area and wait for Special Forces, the Marines grew bored. Guzman and Folsom searched the palace, then watched in amusement as heavily armed SEALs "assaulted" the deserted structure.[20]

It quickly became apparent that the threat to Tikrit and nearby areas was not *fedayeen*, but a large Kurdish force pressing in from the northeast, eager to take revenge on Saddam's home town. The prospect of tangling with the Marines led the Kurds to abandon their campaign, but a very peculiar threat arose. A delegation of elders from Benji, 25 kilometers north of Tikrit, met with the Marines to voice their concerns over prowling American attack and scout helicopters. When the 4th Infantry Division sent an officer to share plans for a major raid on Benji, the Marines instead took him for a tour of the town and lunch with the locals.[21]

The major task of the Marines was to restore order, since the Ba'athist police and security forces had collapsed. The process was hamstrung by the inability to screen applicants for positions on the new police force.

On 21 April the task force left Tikrit, and was disbanded. One of Folsom's last acts was to visit the site where he had killed the man in the culvert. The body was gone, but the place stank of death, and Folsom had nightmares about that day.[22]

First Tank was among the units scheduled for early withdrawal. On 20 April it turned over its responsibilities to the Army's 3rd Battalion—69th Armor, and marched south to Diwaniyah.

On 2 May Task Force TRIPOLI was reconstituted and assigned to screen the Saudi border, reportedly a major transit route for foreign fighters. After a few days the task force was dissolved, as there was no credible intelligence of insurgent activity in the area. Senior leaders foresaw no role for tanks in what was anticipated as a rapid transition to Iraqi self-rule.

All things considered, the large Marine force was shipped home with considerable alacrity. Dan Hayes and his comrades spent idle weeks at the University, then returned to Kuwait. "May third was the day we jumped on the HETTs and drove back [to Kuwait]." He found that "Somebody, I don't know who, maybe MSSG or something, went and picked up all those [abandoned] tanks up, and they were sitting down there in Kuwait." When they opened the tanks, "You could tell it had been kind of looted, stole all the MREs and all that stuff. Nobody took my sandals. They were still sitting there in the bottom of the tank."

When the Army took over the compound where Crabtree's maintenance group was based, "We went down to a dump. It was a filled-in dump. We just laid out there for a while." After several such moves, "We got the word to start rolling back to Kuwait" in small convoys.

In the temporary camps, age-old problems again surfaced. Daniel Benz: "Another big thing I got to point out is field sanitation. I worked very hard to enforce discipline for, here's where we shit. Don't shit anywhere else. I hate to take a negative—I always preface everything. They called me Mr. Caveat. Anyway, a seven-ton drives through the area; he's got his headlights on. What does it do? It illuminates some guy's butt cheeks. He's taking a shit right in our area. I'd just go over and light the guy up: 'You stupid, selfish, –' I didn't talk like that. I don't use profane or abusive language talking to our Marines. But you see that a lot. What happened was a lot of people started getting sick. We called it the 'ass piss' when people vomit and shit at the same time, uncontrollably. I would say probably 80 percent of the battalion got it at one time or another. But the company headquarters group, none of us got sick. Some guys would get sick and then get better and get sick again and they'd get better. We always had a lot of people seriously sick."[23]

Mike Mummey: "We were out of there by Memorial Day. I went back to Twentynine Palms. The rest of the battalion got stuck on the USS *Boxer*—on the prison ship *Boxer*. They didn't get back until July 26. When I got back my job was to turn the battalion back on. . . . We had sealed everything up, turned everything in, shut off water, lights, power." The advance party consisted of selected Marines from each company and those due for transfer or discharge, and replacements had to be integrated.

NOTES
1 Folsom, *The Highway War*, p. 261–262.
2 Ibid, p. 261–262.
3 Anonymous, *Operation Iraqi Freedom Iraq—1st Tank Battalion,*. p.30.
4 West and Smith, *The March Up*, p.172–173.
5 Ibid, p.179–180; Folsom, *The Highway War*, p. 267–275 .
6 Anonymous, *Operation Iraqi Freedom Iraq—1st Tank Battalion.* p.32–33.

7 West and Smith, *The March Up*, p.200–205.

8 Anonymous, *Operation Iraqi Freedom Iraq—1st Tank Battalion*,. p.35.

9 Anonymous, *Operation Iraqi Freedom Iraq—1st Tank Battalion*. p. 39.

10 Smith, *Baghdad and Beyond*.

11 West and Smith, *The March Up*, p.213.

12 Ibid, p 216

13 Ibid, p. 219.

14 Galuzska, *Interview with Major Daniel Benz*, p. 26–27.

15 Folsom, *The Highway War*, p. 291–294.

16 Shinseki's adamant insistence that more troops would be needed for occupation duty was not popular with his civilian superiors, and was a major factor in his dismissal.

17 Folsom, *The Highway War*, p. 303–342.

18 Anonymous, *Operation Iraqi Freedom Iraq—1st Tank Battalion*. p.39–41.

19 Folsom, *The Highway War*, p. 327–339.

20 Ibid, p. 347–348.

21 West and Smith, *The March Up*, p. 251.

22 Folsom, *The Highway War*, p.366–369.

23 Galuzska, *Interview with Major Daniel Benz*, p. 31–32.

CHAPTER NINE

Return to Iraq

He shot at the strong and he slashed at the weak
From the Salween scrub to the Chindwin teak:
He crucified noble, he scarified mean,
He filled old ladies with kerosene:
While over the water the papers cried,
"The patriot fights for his countryside!"
But little they cared for the Native Press,★
[Or] The worn white soldiers in khaki dress. . . .
(★ local militia)
—Rudyard Kipling, "The Ballad
    of Boh Dah Thone," 1888

"NATION BUILDING," THE messy reconstruction of a collapsed society—a failed state in twenty-first century jargon—was scorned by the neo-conservatives who had planned and successfully conducted the lightning war against Iraq. Meantime the Marines had been tasked with nation-building for well over a century. Though they might not welcome the role, they at least had some institutional experience in this most difficult of tasks. Not so their civilian superiors. Their plan was to rapidly stand up a new Iraqi government, and to withdraw from Iraqi cities by September 2003.[1] Dissenting voices like that of Army General Eric Shinseki, who raised the warning that larger forces might be required to exploit the victory so easily won, were none too subtly stifled. The only saving grace was that the various Iraqi factions vented much of their energy in internecine warfare and ethnic cleansing, without mounting a truly coordinated anti-Coalition insurgency.

187

Coalition forces (predominantly American) found themselves in possession of a nation wracked by civil war in all but name. Not only central and local government rule, but even basic services such as drinking water and electricity, had collapsed. Vast stores of weapons and military supplies had fallen into the hands of unrepentant Ba'athist loyalists, international terrorists who had gravitated to this new front against America, fundamentalist religious paramilitaries, tribal militias, and criminal gangs.

The problem was exacerbated by the insistence upon "de-Ba'athification," a purge of any who had served the former regime. No lessons were drawn from experience in the occupation of Nazi Germany or Imperial Japan after their collapse. The insistence upon de-Ba'athification disqualified most of the experienced civil servants, police, and others who had served the former regime only as uncommitted functionaries, but in fact possessed all the experience in operating the diverse and fractious society. Novice police and other units were given cursory training, and expected to stand up to fanatical—and often experienced—*jihadis* who flooded into the country, and equally fanatical military and police of the former regime.

Faced with increasing violence, two American Army divisions scheduled for much-needed relief, and a shortfall of Coalition troops, the Joint Chiefs concluded that the Army and Marine Corps would need to provide at least six fresh battalions each, plus a divisional support infrastructure.[2]

Though Commandant General Michael Hagee and his staff anticipated this new mission, it was hardly welcome. Wars deplete supplies and wear out equipment at a phenomenal rate, and logisticians cannot simply replenish either from the local big-box store. Planners did not anticipate reconstitution of MPS stores prior to early 2004. The new mission would require even more utilization of Reserve manpower, and utilization of stores and equipment from bases on Okinawa and elsewhere.[3]

After several changes during the planning process, the force was settled upon as a division-scale Marine Expeditionary Force (MEF) with the usual aviation assets. The force would be built around the 1st Marine Division and would consist of: Regimental Combat Team (RCT) 1 of three infantry battalions, an Army motorized battalion, and a recon battalion; RCT 7 of three infantry battalions and an LAR Battalion; the Army's 1st Brigade Combat Team, 1st Infantry Division of one Marine and two Army infantry battalions (one Army battalion was converted from armor, and retained a few tanks), and an artillery battalion. Division troops would include a re-serve Marine infantry battalion, an artillery battalion as provisional MPs, and one company each of tanks and AAVs; two batteries of Marine 155mm howitzers would eventually be added. To fully reconstitute this force, the Corps would strip assets from other units.[4]

The mission was unenviable. The I MEF would have to control al-Anbar and northern Babil Provinces, one-third of Iraq's total land area that stretched from the suburbs of Baghdad across a vast expanse of river bottoms, desert and grazing land to the borders of Syria, Jordan, and Saudi Arabia. It was a microcosm of Iraq's problems, with a large Shi'ite population in the east (heavily infiltrated by Iranian agents and with organized religious militias), and a largely Sunni population in the west (with regime loyalists and ties to militant Saudi-based *Wahhabi* fundamentalists). This desperately poor region had been largely bypassed in the original invasion that concentrated on occupying the economic and population centers, leaving old regime units and arms caches largely intact. Through the area ran the "rat lines"— infiltration paths that brought foreign fighters along centuries-old smug-gling routes. With only one overstretched Army cavalry regiment to patrol it, limited contacts led the Army to underestimate resistance. The region had become a *de facto* enemy sanctuary.[5]

The worst trouble spot by far was Fallujah, the "city of mosques." The site had a military history reaching as far back as the writings of Xenophon. More recently a British force savagely quelled a revolt in 1920, at the cost of 1,000 British army and 10,000 Iraqi dead. Even the ruthless hand of Sad-dam had lain lightly on Fallujah, a lawless city whose main industry was smuggling. The residents of the city were already ill-disposed toward the Coalition forces; an errant British bomb had struck a marketplace in the 1991 war, killing many.[6]

The 82nd Airborne Division was the first of several units to be respon-sible for the city. Trained for aggressive military missions, not peacekeeping, there was no nation-wide civil affairs program; such programs were handled division by division, with civil affairs officers seldom having the ear of senior commanders.[7] Assured that a new Iraqi government would be quickly estab-lished, most commanders probably saw little need for such programs.

The first hostile incident occurred on 28 April 2003 when soldiers, confronted with an unruly crowd and the usual celebratory gunfire, fired upon the crowd. Other units rotated through the area but the situation did not improve. The return of the 82nd Airborne in September was marred by a shootout that killed eight policemen, the killing of a Jordanian guard at a nearby hospital, and the killing of a teenage member of a wedding party. The division eventually established a strong civil affairs program, but the insurgency was already well-entrenched. At first the insurgency in and near Fallujah was not a single, simple resistance organization but a devil's brew of hard-core Ba'athists, Abu-Musab al-Zarqawi's *"al-Qaedah in Iraq,"* and outright criminal gangs.[8]

Al-Zarqawi, born Ahmad Fadil al-Khalayla, was a Jordanian alcoholic and petty criminal who turned to radical Islam in prison. After fighting against the Soviets in Afghanistan, he returned to Jordan where he was jailed for advocating overthrow of the monarchy; a violent prisoner but charismatic in the Charlie Manson mode, he developed his following in prison. After release he drifted through Europe, Pakistan, and Afghanistan, where he considered Osama bin-Laden too willing to compromise with the West. Returning home, he was arrested in early 2001 but jumped bail; he was convicted *in absentia* and sentenced to fifteen years, and later to death for murder.

After September 2001 he turned up in Afghanistan, willing to work with bin-Laden against a common foe. Wounded in an American air strike, he fled to Iraq, where he fought against the Kurds. After the fall of Baghdad he orchestrated bomb attacks on Shi'ites in Iraq, and against the West in Spain and Kenya. Kidnapping and beheading, both of Iraqi leaders and foreigners like Nicholas Berg, became the trademark of his followers; some were political, others strictly crimes for profit, and all violated Islamic law. In lawless Fallujah Zarqawi conspired with hard-line clerics (after killing a few moderates) to set up a sort of hard-line Islamic state, though even the other resistance movements sought to distance themselves from his reign of terror. The world had no idea how bad things were to become in the "Islamic city state." Insurgent rule was particularly odious to the Sufi Muslims, who were forbidden to pray by the graves of their ancestors, and all men were required to grow beards. The Marines would later find grisly "slaughterhouses" with mutilated and dismembered bodies, and announcements and documents decreeing death for such diverse crimes as not removing market stalls from sites near the library and (for women) not covering themselves from head to toe.[9]

On the whole the exchange of Marine for Army units went smoothly with the exception of minor ambushes when the Marines made incursions into former sanctuaries, and a brief exchange of fire with Syrian border guards.

The US Army divisions that the Marines replaced had relied upon "heavy sweeps," incursions by large heavily armed forces. The Marines sought to shift the emphasis from confrontation to pacification, though as early as 29 March, 2/9 began small raids into the city to define insurgent strongholds that were the source of attacks on forces outside the city. With little perceived need for armor in the tactical environment, the single Marine tank company was patrolling in Humvees.[10]

The Marines never got the chance to implement their own civil affairs program. Blackwater Security was one of numerous contractors operating in the country; housed aboard a US base outside the city, their detachments often did not coordinate with the US military. On 31 March insurgents ambushed a two-car Blackwater convoy, killing four guards. The bodies of three were mutilated, burned, and two were hung from a bridge in western Fallujah; the fourth was dismembered. The atrocity, in direct contravention of a Muslim tradition of respect for the dead, even the bodies of slain enemies, horrified many citizens of the city.

Though such atrocities were hardly new in the history of irregular warfare, coverage by news crews fanned the flames. Senior Marine officers argued for patience to avoid stoking the fires even higher. Despite the quiet efforts of local Iraqi community leaders who risked Zarqawi's wrath to recover the bodies and defuse the situation, there arose a clamor for vengeance. Some advocated destruction of the city; there was significantly less appetite for destruction among those who would actually have to go into it. On 3 April Army LtGen Ricardo Sanchez, after conferring with President Bush and Secretary of Defense Rumsfeld, ordered Operation VIGILANT RESOLVE, the cordon and forceful occupation of the city.

Fallujah, with a population of some 300,000 persons covers 30 square kilometers, and is a maze of streets and dead-end alleys bounded by multistory masonry buildings with flat roofs. Front and back gardens are typically enclosed by masonry walls, with barred windows and heavy doors to deter intrusion.

The Marines' resources were inadequate for such an operation, but commanders were confident of success provided the city could be effectively cordoned off to prevent the insurgents from receiving support and reinforcements. The operation would follow an established pattern. Forces would cordon off the city, launch raids to reduce strongpoints and reduce the enemy by inflicting casualties, then systematically seize and hold sectors of the city. Engineers constructed an earthen berm around the south side to interdict the flow of traffic in and out of the city, while 2/2 and D Company, 1st LAR established the cordon. In a throwback to old Marine practices a platoon from Captain Michael D. Skaggs' C Company, 1st Tank would be attached to each of the assault battalions, four tanks plus a headquarters tank to 2/1 and a four tank platoon to 1/5. The third platoon was attached to the 7th Marines.[11]

On 6 April Marine incursions from the northwest and southeast quadrants of the city touched off six days of ferocious urban combat. Even before the beginning of VIGILANT RESOLVE the northwestern Jolan District of the city, with its slums and *souk* (market district) was a flashpoint. Aircraft had received and returned fire, setting parts of the district ablaze.

The Marines systematically fought their way into the city, using tank fire to selectively destroy strongpoints. Rules of engagement prohibited the Marines from firing upon mosques or medical facilities, both of which the insurgents used freely as arms depots, command posts, and firing positions.

Nick Popaditch, recently promoted to Gunnery Sergeant, had volunteered to return to Iraq with C Company, 1st Tank and was now a platoon sergeant. On Tuesday, 6 April his two-tank section was supporting F/2/1 in Jolan. The tanks waited in reserve until summoned by the call "Roll tanks, roll tanks." An infantry patrol had been ambushed.[12]

RPGs that were the insurgent's most common anti-tank weapon could not penetrate the composite armor of the tanks, but like most tankers Popaditch was fighting his tank standing in the open hatch for better situational awareness. The main cannon was used against improvised pillboxes built into the ground floors of heavy buildings, but against most targets the heavy machine guns mounted on the roof were more useful and could be aimed and fired more quickly. One insurgent stepped out and fired an RPG that detonated harmlessly against the turret face, and ducked back inside a masonry building. Popaditch simply chewed through the wall with the .50 caliber machine gun. Popaditch later recalled the feeling of power and elation that that he felt, maneuvering the powerful tank in battle. The only real obstacles to the tanks were grids and strands of wire strung from poles, and potential streetside bombs.

After dismounting to confer with an infantry officer, Popaditch decided to reverse the usual practice in street fighting: the less vulnerable tanks would lead the way, while the infantry provided overwatch for snipers, the main threat to the tank commanders and loaders who manned the second turret-roof machine gun.

Thanks to the technology of infrared targeting and aerial observation, the fighting continued until 0400 on the next day, when the tanks pulled back for a brief rest. In the pre-dawn gloom the infantry gunned down three insurgents who tried to creep up on the sleeping tankers.

At dawn the tanks were back at work. A radio message summoned the tanks, and they were soon flushing insurgents down the narrow streets and alleyways. As Popaditch's tank "Bonecrusher" passed the opening to a side alley, he spotted an insurgent taking aim at the tank with an RPG. Assessing the man's calmness as he took aim, Popaditch knew this was no amateur. But the real threat was behind him, out of his field of vision.

The unseen gunner loosed an RPG round that probably struck Popaditch's helmet a glancing blow and detonated. He felt no pain, only a massive blow to the head, a blinding white flash that quickly faded to blackness, and an "electric humming" in his ears. The shaped charge explosion had undoubtedly vented its main jet past Popaditch's head, but had ripped through his helmet, destroyed his right eye, and sent a dime-sized piece of metal sideways through his skull to lodge against the optic nerve behind his left eye. Other shrapnel ripped into the gunner, Corporal Ryan Chambers and the loader, Lance Corporal Alex Hernandez.

Blind and deaf, Popaditch struggled back to his feet in the cupola, oblivious to the chaos around him, unaware that the explosion had set fire to the gear in the gypsy rack. The driver, LCpl Christopher Frias, quickly moved the tank back and began a mad race out of the city, through the twisting streets and over the berm.

A news crew at the assembly area recorded the arrival of the tank— Popaditch upright and bathed in blood, bloodied Chambers vainly trying to extinguish the flames in the gypsy rack with his water bottle. Navy medical corpsmen swarmed over Popaditch. One kept asking annoying questions to which he already knew the answers—among them Popaditch's blood type, which was both on his identity tags and written on his armored vest. In reality the corpsman was trying to assess brain damage and make sure he was not slipping into a coma. Finally Popaditch puked up his most recent meal all over the troublesome corpsman. Proceedings were interrupted by a mortar attack; corpsmen stripped off their own body armor and draped it over Popaditch to protect him. At last, he was anesthetized, waking up on a medical evacuation flight to Germany.

Additional infantry battalions were committed to the city, leading to a throwback to the "bad old days" of Vietnam. Platoons were split into two-tank sections that operated independently, but Skaggs successfully resisted attempts to have tanks operate singly in support of companies. In the narrow streets a lone disabled tank would have been a sitting duck.[13]

The Marines had to relearn some old lessons, and Skaggs later noted that "While most understood mutual support, few understood what it actually looked like. The infantryman's understanding of securing a tank was to remain beside it. This position provided little actual security." For infantry, being too close to the tanks is fraught with danger, since they are "fire magnets," and the protecting infantry was driven to cover. The Marines quickly learned to send a small party of infantry ahead of the tank under protection of the its guns, with most of the infantry following at a small distance to protect the tank from attack directed from the flanks or above. Target designation by infantry was also a problem, since the targeting systems sit ten feet above street level. The tankers learned that a sequence of infantry marking the target with tracer, followed by a confirming stream from the coaxial machine gun before the main gun was fired, worked best.[14]

Fighting in other towns stretched resources to the breaking point. While the media was fixated on Fallujah, the battles to isolate the city, with their own hazards, went unnoticed. On 8 April insurgents ambushed Marines north of the city; as two of Skaggs' tanks maneuvered to support the infantry, one broke through the right side of a small road and slid majestically down into a deep irrigation canal, buried to the top of its hull. Efforts to recover it resulted only in the unit's M88—inexplicably missing some of its critical cables—becoming mired. The M88 was freed by another retriever the next day, but the stricken tank was quickly becoming a liability, with Recon Marines and MPs pinned in place to protect it.

Requests to destroy and abandon the tank were denied: General Mattis was intent on denying the insurgents a propaganda victory if they claimed destruction of the tank. Additional earthmoving equipment from Combat Service Support Company-221failed to free the tank, and its defenders spent a second night immobilized in hostile territory. The following morning the second tank departed; it was too badly needed in the city.

On the third morning the Marines awoke to discover that all the vehicles had sunk into the soft ground overnight. They were laboriously dug out by hand. A tank mechanic, LCpl Robert F. Beard, suggested the time-honored method of laying down a path of tree branches, and the M88 was freed. The damaged canal was flooding the area, so drainage ditches had to be dug. The rescuers settled in for another night under rocket and mortar fire.

The next day involved more digging to level the tank, and construction of a road of logs and gravel across a wheat field. The tracks were cut off, and the tank dragged out. The tank sank one final time, but after twelve previous excavations, the Marines made short work of the final recovery.

"There were tears of joy when the Marines pulled that tank out. It was a happy moment" said CWO2 Walter A. Harris of CSSC-221.[15] It was just one more invisible mission.

The successes of the Marines in Fallujah were destined to be undone by politics; they had scarcely established footholds in the city when the Iraqi Provisional Council prevailed upon Ambassador Paul Bremer III to negotiate a cease fire. Insurgent demands focused on withdrawal of Marine snipers and tanks from the city.[16] The demand was evidence that the insurgents realized the threat posed by the tanks. The Marines were clearly reaching the same conclusion, as B Company, 1st Tank was dispatched to reinforce the assault force, but would not arrive until the end of the month.[17]

The Army's General Abizaid unsuccessfully argued against the truce, but elected to break the bad news to Mattis in person. Mattis was late. His command group had just been ambushed, and he arrived at the meeting spattered with his driver's blood. When Abizaid told Mattis he had to break off the offensive, Mattis exploded. "If you're going to take Vienna, take fucking Vienna!" he paraphrased Napoleon.[18]

Bremer had meantime ill-advisedly declared the Shi'ite leader Moqtada al-Sadr an "outlaw," prompting more outbreaks of fighting, from which most of the painfully crafted Iraqi government forces fled. El-Sadr's Mahdi Army militia in Baghdad and southern Iraq threatened I MEF's supply and communications lines. Having handed the insurgents a propaganda victory, Bremer departed the country. As part of the settlement the Iraqi provisional government created the Fallujah Brigade to control the city.

Dave Banning had transferred over to 1st Marines, and was impressed by General Mattis's interactions with the Iraqi officers who were trying to form the Fallujah Brigade. "He was very cagey. General Mattis is definitely not naïve. He was optimistic but not naïve. General Latif would come in and say, 'The young men are very concerned. They see these tanks.'" There was always this round and round about who was going to do what for who first. Like a good poker player, General Mattis knew when he could give a bit and when he couldn't. He would go ahead and move the tanks back 200 meters, make a big deal about it, but in reality it didn't really matter. We could still whack people just the same. They were never able to produce any weapons that weren't rusty and out of date—clearly not the stuff they were using."[19]

The Fallujah battle had been a public relations disaster. The Marines lost control of the all-important publicity battle, while the insurgents made expert use of both television and computer-generated propaganda. Both the Arab-language and Western media eagerly relayed unverified reports of Marine atrocities and mass casualties of civilians. The insurgents trumpeted their achievement to the world, but drew fatal conclusions. Al-Zarqawi's followers now believed they could defeat the Americans, and failing that could manipulate influential Sunnis to force a favorable negotiated settlement to any future battle.[20]

Zarqawi quickly reneged on the agreement, staging attacks outside the city, and handing over a few decrepit weapons and corroded ammunition.

The thwarted Marines reverted to interdicting the rat lines, sweeping the roads, and training Iraqi forces. The Marines were also diverted to the south, where al-Sadr's Mahdi Army had seized control of an-Najaf and other towns.

The al-Sadr family had provided many distinguished Shi'a clerics, but Moqtada was a ruthless opportunist produced by the ruined society. His great-uncle founded the Dawa movement to counter the Ba'athists and Communists, and was murdered by the Saddam Hussein regime in 1980.

His father—Mohammad Sadiq al-Sadr—was killed on Saddam's orders in 1999.

Moqtada later tried to assume the mantle of his dead father, but the established Shi'a clergy rejected the thirty-year-old firebrand who had not completed his religious studies, and blocked his appointment to the Iraqi Governing Council. He acquired his greatest following among disenfranchised youth in the slums, quickly building his own militia, the Mahdi Army. His movement was not so much religious as personal.

The Mahdi Army clashed with American troops in August 2003, but Moqtada was still primarily engaged in a power struggle within the Shi'ite community. His great foes were the elderly Iranian-born Grand Ayatollah al-Sistani and his adherents, who advocated a parliamentary democracy. In October 2003 a warrant was issued for Moqtada's arrest for complicity in the assassination of the Ayatollah Abd al-Majid al-Khoei, but Ambassador Paul Bremer elected not to enforce the warrant.[21]

Stymied in one power grab, Moqtada's followers flocked to an-Najaf, site of the Imam Ali Shrine.[22] Moqtada's followers allegedly assassinated a leading cleric, seizing control of the physical keys to the shrine.[23]

The terms of a truce negotiated by al-Sistani in May granted the militias so-called exclusion zones around the Imam Ali and Kufu Mosques, which became havens for militia activities. The Mahdi Army was better organized and equipped than the government troops and police, and with Coalition troops excluded from the city they attained effective control. They grew bold enough to raid police stations. One policeman was captured and tortured, the proceedings broadcast live on a government radio frequency.[24]

In June the political situation changed with the entry of a whole new leadership team in Baghdad: Iyad Allawi became Prime Minister of the new government, John Negroponte replaced Paul Bremer as Ambassador, and the new commander of the Multi-National Force became General George W. Casey. Violence between the Iraqi police and the Mahdi Army escalated, and the 11th MEU was moved to the region.

The new 11th MEU, built around 1/4, replaced Army Airborne units, but it was hardly a smooth transition. Army guides led advance Marine detachments into zones that Coalition troops had avoided, leading to confrontations. The MEU's CAAT was fired on while moving into the area, and experienced its first ambush within ten minutes of arrival.[25]

Tensions continued to escalate until a patrol investigating possible weapons storage sites observed militiamen moving heavy weapons in the Sadr family compound east of the Imam Ali Mosque, a flagrant violation of the truce. An accidental shot, probably fired by an Iraqi government soldier, triggered an open battle and the Marines withdrew. In the pre-dawn hours of 5 August the militias launched the first of two attacks on a police station, triggering another battle in which the LAV-25s served as pickets, controlling the open ground of Revolutionary Circle between the mosque and the Sadr family compound. Fighting escalated when a helicopter was shot down; the CAAT and LAVs raced to rescue the crew, and the tank platoon assigned to the MEU was called in. The single casualty of the crash was treated by an Iraqi physician from a nearby clinic, and the crashed helicopter was recovered by an M88 to avoid presenting a propaganda prize to the Mahdi Army.

The MEU commander sought to create a buffer zone to protect civilians—and the beleaguered police—from the Mahdi Army by controlling the sprawling cemetery north of the mosque. Tanks and CAAT Humvees patrolled the adjacent areas, and the LAVs and AAVs supported the infantry by firing into the cemetery. Air Force AC-130 Spectre gunships and vehicles with night vision devices continued to fire into enemy-controlled areas throughout the night.

At 0500hours on 6 August the Marines resumed their assault into the cemetery, a random maze of underground tombs and stone crypts that stood up to three meters above ground level. In places the crypts were so closely packed that Marines could not squeeze through the gaps. The tanks supported the infantry by fire, moving up the winding road that bisected the cemetery.

The brutal heat—115F (46C)—was even greater inside the closed tanks. The leader of the tank platoon's light section succumbed to heat exhaustion and drove his tank to and fro on one of the main traffic arteries "in an erratic manner." The Platoon Leader, Lt. Russell L. Thomas, prevailed upon the MEU commander to allow him to relay his tanks, sending some to the rear to replenish ammunition and allow crews to recuperate. Thomas recalled that athletes sometimes used intravenous fluids to avoid dehydration, and had his Navy medical corpsman administer fluids, often leaving the IV lines in his men's arms. In the following days this would become a feature of the tank fighting, with crewmen having up to three IV lines dangling from their arms, and fluid bags slung from the turret ceiling.[26]

Limited forces eventually accomplished what the Mahdi Army could not; the US Army notified the MEU commander that requested reinforcements would not arrive for days. The Marines broke off the offensive at 0700 on 7 August, withdrawing from the cemetery.

The Iraqi government's negotiation of a suspension of the offensive on 11 August emboldened the militias. The Marines, and the newly-arrived 2nd Battalion, 7th Cavalry and 1st Battalion, 5th Cavalry, continued to launch raids, air strikes, and artillery fire against the militias. Often the raids were ruses: when the militias opened fire on the ground troops, they were immediately inundated by artillery fire.

The newly-arrived Army battalions, with fewer infantry dismounts and new to the urban terrain, suffered most. On 15 August a militiaman leapt atop a tank and fired into the open hatches, killing the tank commander and gunner. The quick-thinking driver threw the tank into reverse and crashed into a building. The building collapse buried the tank, but killed the attacker. After several conferences, Marines were assigned as close protection to some of the Army tanks.[27]

In the close-quarters fighting the Marines learned to take advantage of the tanks' reputation as "bullet magnets." The insurgents quickly learned to fear the distinctive engine drone of the AC-130 gunships, but could not resist the temptation to fire at tanks, revealing their positions.[28]

On 21 August the Marines raided the occupied Kufa police station, on the east side of the city. The tanks, supporting two platoons of infantry, were under orders not to use their main guns as a nearby mosque was in the line of fire. The plan was for the tanks to lead the advance, draw fire, and call air support. Instead, the tanks found themselves under heavy fire and closely-pressed by numerous militiamen, and Lt Thomas feared his small group was about to be overrun. GySgt Philley fired two main gun rounds and struck the mosque.[29]

By late August the Coalition forces leadership and the provisional Iraqi government had grown weary of the militia's intransigence and brutality. In a final assault three American battalions and supporting arms would bear the brunt of the fighting. Four Iraqi infantry battalions and the 36 Commando Battalion would conduct combat operations, but most importantly would conduct the final assault on the Imam Ali Mosque. Polish commandos and various supporting arms would round out the force.

On 24 August tanks led the seizure of a staging area between Routes CORVETTE and CAMARO, parallel broad boulevards that led from the western side of the "ring road" toward the front of the mosque. They were closely followed by engineers to clear obstacles, mines, and IEDs, then infantry mounted in AAVs. Army and Marine tanks blasted away at strongpoints, expending their initial load of ammunition in the first half-hour of combat.

On 25 August the temperature soared to 130F (54.5C) as the soldiers and Marines fought to clear the western approaches, slogging through the built-up area between the boulevards. Tanks served as direct fire artillery. In one instance supporting arms were unable to suppress fire from a strongpoint underneath the overhang of a large building. A single round from a tank brought the entire building crashing down upon the defenders.[30]

By days' end the defenders were painfully aware that the battle could end only one way, and desperately sought a negotiated truce. Unwilling to grant a cease fire that the militias could again turn to their advantage, Coalition leaders and the Iraqi government declined to negotiate, relentlessly pressing the defenders back toward the mosque. On the afternoon of 26 August, with the power of the Mahdi Army conclusively broken, the government negotiated a cease fire.

NOTES

1 Kenneth R. Estes, *US Marine Corps Operations in Iraq, 2003–2005*, p.10. Much of the background information for this chapter is drawn from this source, even where not specifically cited. Note that page references refer to the review manuscript copy, and may not match final pagination.

2 Ibid, p.7.

3 Ibid, p.7–9.

4 Ibid, p.13, 15–16.

5 Ibid, p.14, 26, 35–39.

6 John R. Ballard, *Fighting for Fallujah*, p. 1–4.

7 Vincent L. Foulk, *The Battle for Fallujah*, p. 10–11.

8 John R. Ballard, *Fighting for Fallujah*, p. 4–8.

9 Vincent L. Foulk, *The Battle for Fallujah*, p. 47–61, 154–156, 217–219.

10 John R. Ballard, *Fighting for Fallujah,* p. 12; Kendall D. Gott, *Tanks In The Cities,* p. 94; Major Robert J. Bodisch interview, April 2012.

11 Skaggs, *Tank-Infantry Integration,* p. 41.

12 The account of Poaditch's final action in Fallujah is abridged from Nick Popaditch and Mike Steere, *Once A Marine,* p. 3-15, 274–286.

13 Skaggs, *Tank-Infantry Integration,* p. 41.

14 Ibid, p. 42–43.

15 Valliere *et al, CSSB-1 helms mission to unearth sunken tank in northern Fallujah*

16 John R. Ballard, *Fighting for Fallujah,* p.122.

17 Chang, Tao-Hung, *The Battle of Fallujah: Lessons Learned on Military Operations on Urbanized Terrain (MOUT) in the 21st Centur,* p.33.

18 Cloud, David, *The Fourth Star,* p.153.

19 Lassard, *Interview with Major Dave Banning,* p. 18–19.

20 John R. Ballard, *Fighting for Fallujah,* p.12.

21 Vincent L. Foulk, *The Battle for Fallujah,* p.118.

22 The Imam Ali Shrine is the tomb of Ali, grandson of Mohammed and his last male heir.

23 John R. Ballard, *Fighting for Fallujah,* p.29.

24 Francis X. Kozlowski, *The Battle for An-Najaf,* p. 2. The general outline of the fighting for an-Najaf is drawn from this source.

25 Ibid, p. 3.

26 Ibid, p. 24, 41.

27 Ibid, p. 31.

28 Sattler and Wilson, *Operation AL-FAJR: The Battle of Fallujah—Part II,* p. 12.

29 Francis X. Kozlowski, *The Battle for An-Najaf,* p. 34–35.

30 Ibid, p. 40.

CHAPTER TEN

Tipping Point: The Second
Battle of Fallujah

You can feel a tank coming down the street, you can feel the street rumbling. In An Najaf, that's what caused a lot of them (insurgents) to pull back.—Lt. Col. Bart S. Sloat, Commanding Officer, 1/4[1]

We need to clean out that rat's nest. The longer we wait the stronger they get.—Anonymous Marine Corps officer, August 2004[2]

ROBERT J. BODISCH GREW up as the son of a Vietnam-era Marine, joined the Reserve as an enlisted infantryman in 1991, and was commissioned in 1996. Armor was his first choice out of Officer Basic School, and he joined 1st Tank Battalion in January 1990 as a Platoon Leader in Bravo Company. After attending Intelligence School he was assigned to Third Army in Kuwait in 2002 as a targeting office for Operation ANACONDA in Afghanistan. Greg Poland managed to get him transferred to Ground Operations for the remainder of his tour. He joined Charlie Company, 2nd Tank in early July of 2004 and was told he had three months to train his company before going to Iraq. "Pretty much immediately, within a week or so, I was in the field. . . ." Bodisch also found time to read everything he could find on tank-infantry combat in Hue City, 1968.[3]

There was a huge personnel turnover, but "Luckily for me the Marines I inherited, a pretty good percentage of them had been veterans of the first OIF. I inherited the First Sergeant who had been the First Sergeant and the

Tank Leader, and a lot of the key billet holders were all veterans. . . . That included a couple of my officers."

One of the men Bodisch inherited was Bill Hayes, transferred from Bravo Company. "I remember asking the Master Gunner at the time, 'Man I have got to go. I cannot sit here bored out of my mind. . . .'"

Hayes thought Bodisch was "a smart guy" but "a little rusty with the tank" after his intelligence tours. Still new to the company, Hayes was surprised to be assigned to the CO's tank. "For the CO, you want to make sure his gunner knows what's going on . . . . I want to make life easy for the CO, because he's gonna be pretty busy."

Bodisch had been in contact with Mike Skaggs, never imagining that he would someday replace him. As the result of Skaggs's experience, additional training was fitted in wherever possible.

"While we were at the gunnery range, any kind of 'white space,' any kind of time between actual gunnery training qualifications, I always had my Marines rotate through different training packages. Convoy, live fire courses, get out of vehicles and do fire and movement. Basic infantry tactics. Things that tankers don't like to do. In fact, there was a day out there where I almost had a mutiny by my gunnery sergeants, my platoon sergeants. They were significantly questioning my training program. Luckily my First Sergeant, who was a heavy hammer, was able to nip that in the bud and get those gunnys on board with my plan. I basically had to sell that to them, get them to buy off on it, and they did."

Captain Chris Meyers of Alpha Company had been fortunate enough to get to Iraq and go over the ground with Mike Skaggs. "He just seemed to have a little more prep time for his company. When I got there it was pretty much balls-to-the-wall."

Coalition activities nationwide were curtailed by the force rotation of July–October 2004. Coalition forces slowly regained some measure of control over populous areas, but Fallujah remained a festering sore. Insurgents mounted attacks on Coalition positions in the city, fired upon aircraft, and launched terror campaigns against locals siding with the Coalition.

The insurgents did not waste the time granted them; they built up assets and manpower, and prepared a defense in depth. There were hardened positions inside buildings, fallback positions and weapon caches, and preplanned escape routes. Foreign fighters were pouring in from all quarters.

"I think they had all Mujahedeen [sic] fighters represented except for Filipinos and Indonesians" noted one tank company commander.[4]

The overextended 1st Brigade Combat Team, 1st Cavalry Division was hamstrung by limited civil affairs support and manpower. The vaunted "thunder runs" were completely ineffectual for controlling the insurgency, and inflicted casualties mainly on teenagers who roamed the streets. Daylight incursions by troopers escorting humanitarian assistance convoys were ambushed. All was grist for the insurgent propaganda mills.[5]

When they arrived the Marines initially left the city to the Fallujah Brigade, limiting themselves to occasional forays into the city for meeting with local leaders, and protecting supply convoys that moved along highways through the city. The insurgents felt no such constraints, extending their reign of terror out into the countryside. Through September and October the Marines and insurgents played a deadly game of tit for tat: car bombs and ambushes versus air strikes, artillery and tank fire, and snipers.

In September Bodisch's Charlie Company and Meyers' Alpha Company, 2nd Tank landed at Taqaddam Airfield. Bodisch: "Just on the drive into Fallujah. . . our convoy was receiving small arms fire. That was interesting. Guys who had never seen combat, sitting there in the vehicle that you don't control because somebody else is controlling the convoy. . . . That is definitely a wakeup call. Wide-eyed, looking around and the vehicles are just hauling ass as fast as they can to get through the small arms fire. Right off the bat it was hot. " Only two weeks earlier one of the platoon leaders in the company he was to replace, Lt Andrew K. Stern, was killed by an IED, and both the outgoing and incoming commanders of the RCT Bodisch was to support had been wounded in a rocket attack. In the turnover, the arriving companies simply took over the departing 1st Tank's vehicles.[6]

As part of the week-long turnover, the tank company commander he was to relieve took Bodisch on an orientation ride, with Bodisch as loader. Bodisch was accustomed to the tank commander's view of the battlefield, but

As a loader you don't see anything. All you're doing is taking commands from the tank commander. I was in that situation where I had no idea what was going on around me. I knew he was obviously firing at the enemy and that there were rocket-propelled grenades flying towards us. He was communicating with the regiment;

he was communicating with his adjacent tank platoons, as well as the other tank company and calling in [artillery] fire. I remember thinking as I was playing the role of the loader "Oh my God, I can't believe this guy is this good. He's really, really good."

In combat the loader's position on the company command tank would be occupied by a Reserve artillery forward observer and high school math teacher, Cpl Chris Akridge. Akridge had reported to Parris Island in 1998, the day after his high school graduation because ". . . if I ever got into a combat situation I wanted a Marine next to me." Serving with Mike Battery of the 14th Marines while he attended college, he was activated before his senior year, and arrived in Iraq in September. Trained as a Forward Observer he was the Fires Chief, responsible for coordination with units the battery was supporting. "Captain Bodisch . . . went to my commanding officer, Captain Parsons, and said that 'I need a Forward Observer to go.'" The FOs were already attached to 3/1, so Akridge was the only qualified FO.

"They put me out on the tank farm, and when they were doing maintenance on the tanks, I was doing maintenance on the tanks. They let me set up my own little FO radio to communicate with. I got trained pretty quick." One day on a training range familiarized him with the tank's weapons and systems. "How I could help out if we needed to repair a track. Mostly it was learning which tool to grab if they needed it."

Bill Hayes was promoted to sergeant soon after arrival. As an NCO, "That made that whole deployment about fifty million times better" since he was relieved of things like guard duty.

One of the "enduring missions" the company assumed was rotating the tank platoon at Traffic Control Point One, the most forward position inside the city. "Just about on a daily basis those tanks would be engaged in some sort of firefight. You had a lot of *jihadists* that would constantly test and probe that battle position." It was "the firefight that never ended."

Bodisch: "The snipers were such a problem . . . that Third Battalion, Fifth Marines constructed a concrete tower probably about two and a half stories high. . . . They also had some bunkers adjacent to that tower, and then tanks would come in front of that tower and establish a fixed support-by-fire position." Two weeks later, "A lucky [mortar] shot landed right on top of the turret and injured him [Platoon Leader Lt Joe Buffamante] and his loader. . . . Both Marines would end up getting patched up and refusing to go home, and returned to fight in Fallujah three weeks later."

After an-Najaf, with the region marginally more secure, the Coalition was ready to address the Fallujah problem. The 1st Infantry Division's Operation BATON ROUGE (1–4 October) in Samarra would provide a template, with civil affairs teams moving in close on the heels of combat troops to secure their gains.[7]

Counting on support from Sunnis in parliament, civic leaders in Fallujah proposed yet another permutation of the Fallujah Brigade. On 21 October they presented parliament with their final demands but got no support from the Shi'ites, who had struck a separate power deal with Chaldean Christian factions.[8]

Returning to their roots in the *Manual for Small Wars*, the Marines this time would carefully split the insurgents from their civilian support, and carefully minimize damage to the city. The insurgency was rife with self-serving factions, so intelligence information gleaned from prisoners was used to direct precision attacks, with the additional benefit that the insurgents, suspecting treachery, began to turn on each other.[9]

Mindful of the earlier public relations disaster, the Marines carefully "shaped" the battlefield. Prime Minister Allawi declared martial law. Water supplies were severed. Leaflets and loudspeakers urged non-combatants to leave or at least seek shelter; an estimated seventy-five to ninety percent of the population departed.[10] Media access would be carefully managed.

In Operation PHANTOM FURY, soon to be re-designated AL-FAJR,[11] Army units would cordon the city. Two Marine RCTs would sweep from north to south through the city. Six of the more reliable Iraqi battalions would enter alongside the Marines, policing the cleared sectors of the city. This time air and artillery support would be overwhelming, though carefully coordinated to minimize damage.

Because the Coalition had to telegraph its intent in order to evacuate civilians, US Army and Iraqi units would create a noisy diversion to draw insurgents toward the south side; the main assault would come from the north. RCT-1, consisting of 3/1 and 5/3, C/2nd Tank, the Army's 2/7 Armored Cavalry and an Iraqi battalion would seize the western part of the city. The RCT-1 commander, Col Michael Shupp, a former amtrac officer and LAR battalion commander, had a better understanding of mechanized units than most, and arranged brief training exercises with two of his infantry battalions.[12]

RCT-7, consisting of 1/8 and 1/3, reinforced by the Army's 2/2 Mech Infantry and two Iraqi battalions would attack the eastern half of the city. Forces would advance north to south, aligning at Phase Line FRAN, a broad east-west boulevard where Route 10 passed through the city. Other Iraqi battalions would enter the culturally sensitive mosques. Upon reaching Phase Line JENA, the southern margin of the city, the forces would wheel about and sweep the city from south to north.[13]

Bodisch's Headquarters Platoon and a tank platoon made an observation trip to the north side of the city near Jolan Cemetery, unescorted by infantry, and with a colonel unknown to Hayes riding as loader in the XO's tank. Akridge: "I had a pre-planned target set up for a smoke mission in case we started getting hit, and we needed to get out of there." Hayes: "This was my first exposure to really getting up close to the city. These guys just came out of the woodwork; started running around, lobbing mortars at us, you'd see them running around with their AKs. It was kind of surreal, like 'What's going on here?' These guys are really mad." He recalled thinking how strange it was to see insurgents using grave markers as protection.

Akridge called in a fire mission, and ". . . they were spot-on. They started rolling up in these little cars and doing somersaults out, shooting at us. . . . I was sitting up outside the loader's hatch with my binoculars, trying to make sure the smoke mission came on right. One of the mortar rounds hit near us, the sand blasted our face." The tanks loosed off a few main gun rounds. Hayes: "We left, and that was that. Kind of like for the senior leadership to figure what was going on."

At 2207hours on 7 November the Iraqi 36 Commando Battalion and a task force from 3rd LAR swooped into the bend of the Euphrates River west of the main city, occupying the Fallujah Hospital to prevent fighters from occupying it. The assault force found evidence it was already used as a command center. The hospital would continue to serve as an evacuation center for civilian casualties. The move garnered grudging praise from the media.

Third LAR also seized the bridges and established checkpoints. Boat units patrolled the river, and the Army's Blackjack Brigade moved into positions south and east of the city, closing the cordon. RCTs One (supported by Bodisch) and Seven (supported by Meyers) moved to assault positions.

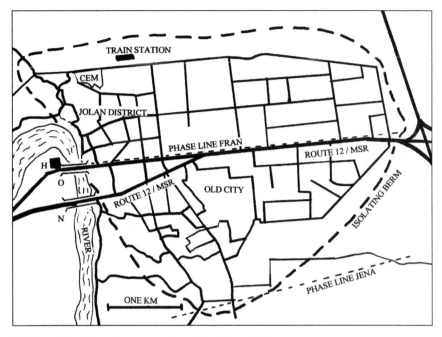

MAP 4. Fallujah and environs. O=Old ("Green") Bridge;
N=New Bridge; H=Hospital; CEM=Jolan Cemetery.

Akridge "did a couple of preparation hits in areas we thought the enemy might be." Hayes: "They had so many concrete barriers" used as obstacles that "we spent most of the night driving around kind of blowing those up."

Engineers severed the city's electrical supply, and aircraft blew breaches through the isolating berm with large bombs. The 1st Marines moved in to seize the Shaklawyiah Apartments, high rises that overlooked the Jolan District. The Marines went door to door, paying the residents $200 to "sublet" their apartments for the duration of the battle. The operation drew heavy mortar and rocket fire.[14]

Late in the day the Marines seized the train station, jumping off point for the main assaults. At 1926 hours 3/5 moved through their breach point to join 3/1 and Bodisch's tanks.

About an hour after entering the city "Our driver was [shouting] RPG! RPG! . . . There was this old guy he had like the robe on, not like a twenty-year old insurgent." Hayes saw him "just as he lowers this RPG, and runs back behind the building." Hayes was startled. "What was that all about?"

Hayes was still confused, because "My advice to anyone who's ever in a war, if you're an infantryman, if your enemy shows up with tanks, just get the hell out of there. Don't fight them, don't shoot at them, don't let them know where you are. Run!

"Does this guy not have any idea what this tank can do to him, and he's just gonna lob an RPG at it and hope to get lucky? He's a dead man walking."

By the morning of 9 November the Marines had secured the expansive Jolan Cemetery and Iraqi Special Forces secured the Hadrah Mosque. This rapid penetration into what the insurgents perceived as their main position of strength and final defensive area would profoundly affect the rest of the battle. The Coalition had achieved a tactical coup, unhinging the enemy's battle plan from the very beginning.[15]

Unhinged or not, the insurgents could still bite back. The assembly area used by Bodisch's Charlie Company tanks was hit by a rocket. "The vehicle next to me, which was an amtrac, got hit and killed a Marine and gravely wounded a couple of other Marines from that vehicle." Bodisch admired the Reuters photographer attached to his company, and thought she had taken "incredible pictures." But when the amtrac was hit "She was right there to see it, and I think she got shocked by it and ended up begging to leave."[16]

With utilities and cell phone services interrupted, the insurgents had trouble mounting coordinated attacks, despite using mosque speaker systems for communications. White surrender flags were common as ruses.[17] There were many such violations of the Rules of Land Warfare.

The Army units used their heavy armored firepower to advance quickly along the streets that constituted axes of the attack, preventing the insurgents from mounting an organized defense, but had too few infantry dismounts to clear buildings. Marines and the best of the Iraqi battalions swept in as a broad wedge behind the armored shock forces, clearing buildings and reducing strongpoints. The other Iraqi battalions followed in train, searching the warren of buildings for bypassed insurgents. In their sectors the Marines would establish contact, and then back off to wait for tanks, artillery, and air strikes to reduce strong positions. In the process "Bulldozers and tanks had turned entire square kilometers into rubble."[18]

Bodisch witnessed firsthand the utility of the huge dozers in urban combat:

There were young lance corporals driving D7 dozers. We took a lesson from the Israelis to bring dozers into the cities to reduce the strongpoints that my tank main gun rounds or the infantry would have a hard time clearing. One of the more interesting observations I had was when a D7 operator was reducing a building that my wingman and I had shot six main gun rounds into. The infantrymen had gone in to clear it, but they were still being fired upon from the rubble and they couldn't go in there to clear it. I remember Lieutenant Colonel Buhl from 3rd Battalion, 1st Marines called that dozer and said, "Hey. Go up there and reduce this thing. Just take it down." The dozer operator went in there and started chopping away at the building with the big blade of the dozer and just brought it down on the house. As he did that, we just watched what was going on. It was another surreal scene. A couple of guys just popped out of the rubble. They literally popped out of the rubble, gray from the rubble and dust, and they started to shoot their AKs right at him, point blank. He wasn't armed because he was in a dozer operating a piece of machinery. We could see the glass and it must have been ballistic because it was taking hits and it was spidering. He had a split second to make a decision: either he was going to get the heck out of there or he was going to do something about it. He turned his dozer towards the enemy, raised the blade and just dropped it on top of the insurgent. I don't mean to be gory but it was weird to see a human being just pancaked under a blade. He took the threat out and luckily the infantrymen were able to go in there and finish the clearing and didn't have any injuries, so he did his job. I had captured that observation and sent it to the regiment and that young lance corporal ended up getting a Bronze Star with 'V' for his actions that day.[19]

This tactic was in response to the inevitable casualties of urban clearing operations. Hayes: "Guys were getting killed left and right running into these houses. I *never* heard casualty reports coming over the radio like that first week I was in there. They were having guys get wiped out left and right, 'Two KIA, three wounded,' Then a few hours later, 'We got three KIA, two wounded.' It was like 'Wow! This is really not good!'"

The infantry eventually began to decline rushing buildings until they had been 'prepped' by the tanks. Hayes: "We would just pull up to it, put about three tank rounds into it and then the D-9s would bulldoze it down."

The dense urban terrain highlighted an age-old problem for tanks; the tanks could not sufficiently elevate their 120mm main guns to fire at positions on the upper floors. Although Hayes—a gunner—disagrees, some sources state that the tanks would typically operate in two "waves," one group leading to draw fire and engage positions in the ground floors and basements, while following vehicles held back to fire into higher floors.[20]

A special round, the HEAT-OR-T developed by the Army for urban combat, proved less effective than the older HEAT rounds. The lack of a canister round, which had proven so useful against enemy ground troops from World War I through Vietnam, was also sorely felt. Tanks were firing at ranges less than the safety-arming range (30-40m) of the main gun rounds. "A lot of the ways we were getting kills from those rounds was from the over pressure of the round itself coming out of the gun tube and the shock from the round itself," said Bodisch.[21]

Akridge: "We mostly used a lot of mortars, because our friendlies were so close to where we were shooting. The one-five-fives would have been a lot. The infantry forward observers, First Sergeant Bayne and Lance Corporal Brooks, they used Mike Battery quite a lot."

Since he was not a tanker, Akridge was more aware that it was "hot as hell" inside the tank. The first three or four days when we had to sleep, all of us, in the tank. That sucked. I guess that was what life in a tank was all about." One day he offered some dried apricots to the other crewmen, and "Their eyes got all big." He had violated the longest-standing tanker superstition. "You can't have apricots on the tank. My grandma sent me a package, so the gunner (Sergeant Mattingley) on the tank that was our wingman, he threw them away for me."

Bodisch found that his new tanks were equipped with MDACT (Mounted Digital Automatic Communications Terminal), a Windows™-based system using EPLRS (Enhanced Position Location and Reporting System); SIPRNet (Secure Internet Protocol Routing Network) allowed communications that were more reliable than the line-of-sight radios, and near-instantaneous over huge distances. "I was actually chatting with my colonel back in Camp Lejeune any time I had a break. I could use chat to communicate to the regiment and some of the battalions. It was very useful because a lot of those radio nets were just clobbered. It's already hard enough to be on the radio with the regiment and all of the battalions and then throw the combat factor on top of that. The radio was confusing to use so having that computer with chat on it was pretty amazing and almost surreal." Bodisch would later find that the system worked even when normal radios were out of range.[22]

Second Lt Jeffrey T. Lee, in command of 3rd Platoon, A Company 2nd Tank, supported 1/8 through numerous heavy firefights in the city. Shot through the arm, he declined evacuation until the Marines had reached their assigned objective, then remained in the open to help drive back insurgent counterattacks. Jeffrey was recommended for a Silver Star for his actions.[23]

This time the public relations battle was far better-managed. With too few Iraqi Muslims to clear the many mosques in the city, the Marines often let Iraqi troops enter first, and then cleared the mosques themselves; the media "seemed not to care that much." A clamor by Sunni voices fell upon deaf Shi'ite ears. The Grand Ayatollah al-Sistani pondered the pleas, and waited until near the end of the battle to call for peace.[24]

Despite heavy fighting, the sweep through the city proceeded ahead of schedule: by late afternoon 2/7th Cavalry linked up with 3rd LAR at the bridges spanning the Euphrates. By 2200hours the cavalrymen had secured Jolan Park, and reached Phase Line FRAN.

Most of the fighting was at ranges far shorter than the Marines had trained for. Bodisch said that "A lot of my engagements were main gun rounds against people targets within thirty feet of the tank." His gunner, Hayes, said that in training ". . . it's always like marksmanship, tank gunnery, can you hit that moving target at fifteen hundred meters? Over there during the Fallujah battle I don't know if . . . any of the tank gunnery skills were ever really put to the test. Especially when you're putting an MPAT round through a house, and the tank barrel is literally ten feet from the door."

At 0100hours on 10 November 1/8, supported by tanks, LAVs, and AAVs, attacked south from the Hadrah Mosque toward Phase Line FRAN. By noon they had seized the Government Center administrative complex, but spent the remainder of the day rooting out die-hard snipers. Marines being Marines, psychological warfare units celebrated The Birthday by playing the Marines' Hymn at top volume through loudspeakers.

An assessment on 11 November suggested that the quick progress necessitated a change in plan; spearheaded by the Army cavalrymen, the Marines continued to attack due south through the city, and by afternoon had crossed Phase Line FRAN. The southeast corner of the city had the most defensive positions. Most faced south and were taken from the rear.[25]

By 12 November Coalition forces controlled 80 percent of the city, but at a cost.

Bodisch's personal log recorded that they had cleared as far as MSR MICHIGAN, an east-west street that divided the city into halves. The tanks were leading the entire RCT-1 advance, and the Headquarters Section tanks, preceded by two other tanks from Lt Joe Buffamante's platoon, headed east and then south on Phase Line HENRY. There they encountered Army vehicles for the first time. Bodisch: "Because we were the Marine tank company, we were the only tank company that would get broken down and attached to the various infantry battalions. Two-seven Cav, they stayed organic: they didn't break their formations below their battalion. It was interesting . . . to see the stark differences between how a Marine tank crew fights, and an Army tank crew fights. . . .

"We don't have many tanks in the Marine Corps, period, so we're willing to operate, echelon, at the section level. The Army never operates below the platoon level. . . . The other key difference is the fact that our Marine tanks were the ones that did all the detailed clearing with our rifle companies. The Two-Seven Cav never broke their formation . . . and their mission was more of a 'thunder run', the kind of tactic that they had perfected in Baghdad and an-Najaf, some of those areas before coming to Fallujah."

Bodisch thought the tactic was effective for short-term urban operations, but less effective for longer, clearing operations. The Army's most important achievement was that "They were able to seize the most important north-south line of communication, Phase Line HENRY, right through the middle of the city." This provided a reasonably secure supply route. "Every night [when] we went firm, I would send my tanks back through that route . . . to get back to the rear and get refueled and whatnot. The drawback was that because they are basically outposting one road those tanks are not in play. They're not providing combat power. They're just doing guard duty."

Another doctrinal difference was that the Army vehicles operated buttoned-up. "Our situational awareness is out eyeballs. Every time we went though that line of communications outposted by those M1A2 tanks and Bradleys, you didn't see their turret moving, you didn't see them moving. They almost seemed like 'Well, just a bunch of tanks out here, and I'm not sure what they're doing.'

"But it did make us feel safer, that they were out there holding that road."

The insurgents made little use of IEDs, but the Marines shredded any abandoned vehicles, just in case.

Just after turning south on HENRY "All hell broke loose. His (Buffamante's) tank got struck by sniper fire, and his loader, who was also the same kid that got injured when that mortar round hit on top of the turret, he got shot right in the hand. It was the only part of his body that was exposed . . . holding onto the slip-ring, with his left hand while he was inside the turret. . . ."

Akridge: "I heard on the radio we're going through a kill zone. I had the loader's hatch open. It was so hot we usually kept it open. I reached up to pull it down. Before I could latch it we had three RPGs hit us right off the bat. . . . It blew the hatch back open, so I had to get out there and close it back down.

"Something happened to the hydraulics or something on the turret, we couldn't really turn as much as we could have. We lost communication or sight with our wing tank. Kinda scary. I thought that time we might have been done for. Sergeant Hayes got a big goofy grin on his face, and we started getting the tank [turret] to turn, and we started loading main gun rounds."

Hayes: "Our wingman tank—the XO tank—was behind us and he was like 'You guys just got hit.' And it was like 'Hit by what?'" Smithley replied that it was an RPG, and said they'd been hit again.

Bodisch: "That was the day my tank got struck by three RPGs, one of which wedged on the side of my skirt. The actual round was still stuck in the side of the skirt. When we were done at the end of the day, I didn't even know it was still there 'til one of the Marines at the refuel point pointed it out."

He recalled that "[Smithley's] tank had also received, just a shit-ton of RPG fire. I could see that this tank that was immediately in front of me, I could see the holes getting punched in the side of the sponson boxes. . . .

They were receiving with so much fire, and they're dealing with a casualty that me and my XO ended up taking the lead. As we were attacking south I was in the lead tank, and Lieutenant Smithley, my wingman, XO, was right behind me. I used to joke with Smithley all the time that I'm getting' struck by all these RPG teams. I'm not able to take them out because they're attacking immediately on my flank. He's able to pick these guys off when they reveal their positions with muzzle flashes from their RPGs.

"He's just whaling away. He was definitely a machine of death, if you will, going down that road. I was drawing all the fire and he was getting all the glory, getting all the killing."

Monitoring Kilo 3/1's radio net, Bodisch heard that an attached team of SEAL snipers riding in a Humvee had also been hit. "The logical guess was that the same sniper had engaged the tank loader and the two snipers. . . ." It was an action fraught with all the confusion and hazards of urban combat.

"I could see two insurgents in an apartment complex . . . two hundred meters to my southeast. I went ahead and requested permission to fire. It took a long time to get the clearance due to being in a cross-boundary situation. I basically grew impatient because a Marine had been struck, the sniper was still engaging. I could see the flash, we could see the bodies, in the thermals, of the insurgents. I could clear my own fires because I knew we were already ahead of the adjacent regiment, and I went ahead and engaged the target."

Hayes thought that viewing the battle through his optics was ". . . like an out-of-body experience," and likened it to a video game. Hayes recalled another long-range engagement, pouring main gun rounds into a high-rise. Bodisch requested an air strike, only to be told no assets were available. Hayes: "We put a few tank rounds into it, and sprayed it with machine gun. The CO [says] 'You see anything, just squeeze that trigger'. We had a tank round in the thing. I'm watching this hole in the side of the building that I just put in there. All of a sudden you see this movement; I remember seeing this guy run from the right to the left. . . . As soon as he exposed himself, it was just pull the trigger. All of a sudden that's smoking hole."

A much-reported incident on 13 November brought home to Bodisch that rapid global communications were a mixed blessing. Inside a small mosque a reporter filmed a Marine rifleman killing an insurgent who he believed was feigning death, a common insurgent tactic. The Reuters reporter embedded with Bodisch's company (the partner of the shell-shocked photographer) could not go into the city, but asked a Marine (who also had not been into the city) for a reaction, and the Marine replied "I would have done the same thing." Bodisch found himself called on the carpet by long-distance telephone from the States for allowing his men to comment. The incident—and others—led to the removal of the reporter.[26]

Near the southern margin of the city Bodisch was supporting India 3/1. "My left flank, to the east, was RCT Seven. I could see the tank section in that fight. . . . He was so close we could just basically shout at each other. That was Lieutenant Jeff Lee. . . ."

The two officers talked face-to-face, confirming the attack plan. At 0730 Bodisch started south, but "I noticed his tanks didn't roll with me. . . . I'm yelling at him 'Well let's go!' Probably said some profane words to him. He looked at me and said he couldn't roll." Lee had not been given the clearance by his supported unit.

"At that point I knew we weren't going to stop. I now had this exposed left flank I'm very concerned about it." This raised real problems with coordinating fires. "Sure enough, right as that's happening I get a call from my XO. He says 'Hey, sir, I don't know what happened, but the infantry Marines, they just stopped.'

"We were going pretty slow, definitely creeping along, but in an urban environment you go fifty feet, that's a horrible distance to go without your protective infantry envelope. At that point I was getting really nervous, because now it's me and my wingman kind of alone and unafraid. We've got nobody else.

"Right as I'm realizing this feeling, I look to my left and I see a group of about thirty insurgents. . . . This is no doubt this is the last stand of these guys." Many surviving enemy had been cornered, with their backs to open terrain outside the city. "They're just now realizing that there's tanks creeping toward them. . . . They were no more than a hundred feet away from me, and they're raising their rifles, RPGs, some of them are running into other buildings. It was just a moment that was seared into my mind. It was surreal. I hadn't seen a formation of the enemy that big up to that point.

"I'm trying to yell to my gunner 'Traverse left! Traverse left!' In that urban environment you're so used to making small traverses that he didn't understand that I needed him to traverse hard left. At that point I just took the TC override, and I'm moving the turret as fast as I can to my left." Bodisch dropped down into the turret to use his own sight extension. "At that point I'm just letting long bursts of coax into this formation of people."

Certain details of the chaos were unforgettable. "The first guy that I had hit with that burst of ammo, it looked like I had hit him right in the leg with one round, and that guy dropped immediately. Dead cold, with one round to the leg. The second guy I put probably a good fifteen round burst into his torso, and yet he was still able to somewhat stand up, and he was yelling at me, shaking his weapon before he hit the ground. It's kind of interesting to me how the human body reacted so differently to us returning fire."

Bodisch was surprised that "The enemy wasn't running away. The enemy in fact reinforced themselves, and started charging toward the tank. We pushed the main gun at this point, and we're just blasting away with main gun, coax, and fifty-cal. It was like a shooting gallery. They were able to close the distance between shooting and loading a main gun round and bringing small arms fire on them. But we were able to pretty much obliterate every one of them, except for ones that had managed to get away."

Akridge was in the turret sorting out the location when "Captain Bodisch jumps down and starts cussing and I realize we're under attack. . . ." Bodisch: "We were able to call a fire mission on the area where we thought the insurgents had displaced. They actually went just on the other side of the southernmost wall of the city." Akridge "They went behind this berm, so I called in an air-burst type thing. . . . At the same time, while I was coordinating the fire, I was also loading the tank main gun rounds because they were all over the place. It was one of those things . . . it was just perfect." The fire mission was filled by Army heavy mortars; the first round was on target. Bodisch: ". . . we just did a fire for effect, and we just kept repeating it and repeating it. Third Battalion, First Marines later on that day would find twelve bodies on the other side of that wall."

On 15 November the various forces aligned on Phase Line JENA and prepared to turn about for the final sweep through the city.

On 16 November the city was declared secure, though sporadic firefights erupted through 20 November. Patrols scoured the city. Much of the resistance seemed to be encountered in the eastern half of the city, where fighting with air strikes and tank support continued through 11 December. Hayes: "From mid-November to mid-December we did a lot of the back-clearing. They were still finding guys. . . . In combat it's always the intense boredom, and then you get the call over the radio, we've been clearing houses and then the twentieth house, there's twenty guys in there."

Akridge was returned to his artillery battery job, replaced by a communications specialist.

Detailed search operations later revealed numerous weapons caches in mosques, crude chemical weapons labs, and manuals for preparing anthrax. The Marines organized tours of these discoveries for the media, and the press videotaped welcoming civilians.[27]

In late December civilians who had fled the city were allowed to return, though Marine patrols were still encountering occasional resistance. Civil affairs teams moved in to help make reparations for battle damage, and to restore utilities and infrastructure as quickly as possible, all a part of "Phase IV" of the operation—"transition to an interim emergency government." Reconstruction was hampered by the recovery of the dead. Though the Coalition forces had established joint mortuary affairs teams to recover and inter dead insurgents according to Islamic practices, the insurgents had no qualms about booby-trapping their own dead, a desecration contrary to Islamic law.[28]

The operation was an overall success, though Abu Musab Al-Zarqawi and most of his foreign fighters fled prior to the battle, leaving local fighters to face the offensive. Efforts to reconstitute the insurgency within the city were at first unsuccessful, given stringent local security procedures. Though the battle significantly weakened the insurgency in this Sunni stronghold, an immediate effect of the death and destruction was to strengthen Sunni political intransigence. They boycotted the January elections, a move which backfired when Shia factions gained increased power in the government.

The brief but violent battle for Fallujah had upset some long held beliefs about the utility of tanks in urban battle. Since World War II the US military had stressed avoiding urban warfare. For some, the Russian debacle in Chechnya reinforced that maxim. In Fallujah the armored vehicles had proven invaluable. Protected by sufficient infantry, they provided enormous firepower to reduce strongpoints, and were far less vulnerable to IEDs than infantry in the narrow streets.[29]

Evaluations of the second Fallujah battle were and still are mixed. In retrospect, the battle was a clear moral victory, eliminating the odious "city state" created by al-Zarqawi's adherents. They had failed to inflict massive casualties and thereby salvage strategic victory from tactical defeat by convincing the American public that the war was unwinnable, as the Communists had done with the Tet Offensive of 1968. But tenacious insurgents immediately began to flow back into the city with the returning citizens. There was still a long, long way to go.

For the tanks, the lesson was that specialized protection was needed for urban combat; the Army developed the Tank Urban Survival Kit (TUSK) to provide better protection for crewmen using turret-top machine guns, and better electronic imaging.

NOTES

1 As quoted in Jim Goodwin, *Marine tank unit crucial to stabilization of Iraq*, unpaginated.

2 Foulk, *The Battle for Fallujah*, p. 124.

3 Fike, *Interview with MAJ Robert Bodisch*, p. 3 -5.

4 Ibid, p. 11.

5 Foulk, *The Battle for Fallujah*, p. 126–128.

6 Author, Major Robert Bodisch interview, April 2012.

7 Foulk, *The Battle for Fallujah*, p. 151.

8 Ibid, p. 205–207.

9 Sattler and Wilson, *Operation AL-FAJR: The Battle of Fallujah—Part II*, p. 14.

10 Foulk, *The Battle for Fallujah*, p. 210–211; Gott, *Tanks In The Cities*, p. 95.

11 Operation al-Fajr is called al-Fahr in some sources. The word mean "Dawn."

12 Fike, *Interview with MAJ Robert Bodisch*, p. 5.

13 Gott, *Tanks In The Cities*, p. 95–98.

14 Ballard, *Fighting for Fallujah*, p.55–56.

15 Estes, *US Marine Corps Operations in Iraq, 2003–2005*, p.91.

16 Fike, *Interview with MAJ Robert Bodisch*, p. 7–8.

17 Foulk, *The Battle for Fallujah*, p. 214.

18 Ballard, *Fighting for Fallujah,* p.56.

19 Fike, *Interview with MAJ Robert Bodisch*, p. 10–11.

20 Gott, *Tanks In The Cities*, p. 98; Ackerman, *Relearning Storm Troop Tactics: The Battle for Fallujah*, p. 54.

21 Fike, *Interview with MAJ Robert Bodisch*, p. 6.

22 Ibid, p. 6–7.

23 Estes, *US Marine Corps Operations in Iraq, 2003–2005*, p.83.

24 Foulk, *The Battle for Fallujah*, p. 215–217.

25 Ballard, *Fighting for Fallujah,* p.64.

26 Fike, *Interview with MAJ Robert Bodisch*, p 7.

27 Ballard, *Fighting for Fallujah*, p.77; Foulk, *The Battle for Fallujah*, p. 219.

28 Sattler and Wilson, *Operation AL-FAJR: The Battle of Fallujah—Part II,* p. 23.

29 Ballard, *Fighting for Fallujah,* p.101–104.

CHAPTER ELEVEN

Harrying the Insurgency

*I come in peace. I didn't bring artillery. But I'm pleading
with you, with tears in my eyes: If you fuck with me,
I'll kill you all.*—General James Mattis, USMC

IN THE TRANSITION TO counterinsurgency (COIN) operations, the Marines coordinated with the US Army, newly raised
Iraqi formations, and troops of the British Black Watch to
launch pursuit and exploitation operations intended to keep the insurgents
off-balance and on the run. Towns surrounding Fallujah were cordoned
and swept, even as the effort to rebuild Fallujah commenced.

Inside the city the Marines still conducted sporadic assaults on insurgent strongholds, and the tanks were often called upon to help with precision fire. At the first opportunity to gather his company together, Bodisch
and his First Sergeant ". . . kind of talked them up a bit. We said 'Things
have to change. You're not killing the first person you see now. That stuff
is over. Now we have to shift back to helping these people.' I was actually
pretty proud that those guys were able to make that shift so quickly."[1]

Hayes recalled a brief period of doing "Classic Marine stuff: trying to
look busy." Then the CO of 3/8 gave Bodisch a section of amtracs, a rifle
platoon from India Company, an engineer section, and occasionally a
Human Exploitation Team and translator, to establish a roving mechanized
force. "We were doing everything the infantry guys were doing. We were
doing cordon-and-knock, we were doing snap vehicle control points, and
we were doing mechanized patrols. In the four weeks that we operated like
that, we uncovered more weapons caches than any of his rifle companies
did. They were new, just getting there, trying to figure things out. My guys
were pretty seasoned by that point."

LtCol S.M. Neary's instructions were simple: "'I want you to go in my area of operations, and I want you to run amok. I want you to do what you think you need to do. I want you to uncover weapon caches. I want you to get into the populated areas, search these homes, search for insurgents, and develop some intelligence.' And that's exactly what we did. It was probably the best mission a tank company commander could have in that COIN environment."

In general, there was little fighting: a few small IEDs, and one night Smithley's tank hit a large mine. Hayes recalled only one unusual incident, when yet another road edge collapsed under the tank. Looking out through his optics at a rural road, Hayes thought "This is not a good idea. That stupid canal just starts to give way, and . . . the tank began to slide to the right. I sprinted out of that tank. Before it even settled, I was out. Across the loader and got out of the tank and jumped. I knew the second that things started to turn 'I'm getting the hell out of here. Fast.'" The tank settled with a forty-five degree list, and partially threw a track, blocking the narrow road. The tank had to be extricated by cabling to four amtracs. The road continued to collapse, and Smithley's tank also had to be extricated.

In April the company's deployment was over. Bodisch: "Captain Matt Youngblood came in to replace me, from Second Tank, and for his seven months he basically did a lot of those type operations. . . . Basically every tank company commander after me did pretty much the COIN-type operations."

All these efforts were geared toward both supporting the 30 January 2005 elections, and preventing the enemy from exploiting the relief-in-place of I MEF by II MEF in the first quarter of 2005. The rotation and scale-down of forces would bring 2nd LAR Battalion, and a number of small mechanized units to be co-located with 1/8 at Camp Fallujah: B Company, 2nd Tank, an AT Platoon from 2nd Tank, Scout Platoon of 8th Tank, and B Company, 2nd AAV Battalion.[2]

John Polidoro, the son of a university professor and a schoolteacher, attended Annapolis "because it was free and it was hard to get into," then chose the Marine Corps commissioning path because "it kind of suited my personality." Though a tank officer by training, most of his combat assignments were with the LAR units.

Polidoro had joined 2nd LAR as the Executive Officer in June 2004:

. . .we spent the next seven months hearing "You're going. You're not going. You're going. You're not going." It wasn't until January 2005 that we actually got notification of our deployment in March.[3]

Pre-deployment training included gunnery, Military Operations Other Than War, urban operations, and a three-week combined arms exercise at Twentynine Palms, but none of the training was specifically directed toward the special conditions in Iraq. Cultural training got lost in the pre-deployment rush. Polidoro continued:

> I think we were focused too much on the conventional fight and didn't spend enough time preparing for the cultural differences between the Western mind and the Arab mind. We could have done a lot better interacting with people if we'd had more training in that. We could have fought conventionally against anybody in the world at that time because we were well prepared, but we didn't end up doing that. We were a little bit off the mark. . . .[4]

The real problems of the rotation came not from the insurgents, but from wear and tear on vehicles never designed for months of continuous campaigning in a brutal environment. In January 2nd AAV Battalion sent a reconstitution detachment to evaluate and refurbish—where possible—the vehicles in place; half of the 84 vehicles in country had to be replaced with vehicles shipped from Camp Lejeune. The following month the even more badly worn tanks and VTRs were replaced by vehicles from MPS stores. Some units like 2nd LAR brought their own equipment from the US.

With the elimination of the major enemy stronghold in Fallujah, the mission evolved to a pacification of the countryside, counter-sabotage, and interdiction of the ever-present rat lines. Where feasible, tanks and AAVs supported infantry sweeps to clear insurgent strongholds, and either kill insurgent leaders or force them to relocate.

Polidoro found that the mission was multi-fold: "We had a counterinsurgency role, a route security role in western Iraq from just east of Haditha out to the borders of Tribil, down in Jordan, and Walid up in Syria. So we basically had about ten thousand square miles to cover, and about seven hundred miles of road that we were continually patrolling, and then some

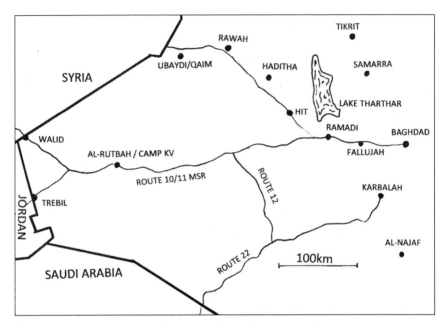

MAP 5. Marine Corps Western Iraq Operational Area. Parallel highways Routes 10 and 11 parallel the Euphrates River; the meandering river course is omitted for clarity.

major urban centers, the town of Akashat (in NW al-Anbar Province), and Riphah, which is about thirty-five thousand or so people. . . . We were doing both clearing operations, targeted raids, and constant patrolling and convoy escort as well."

Convoy operations were a continuing burden, particularly given the insurgents' increasing utilization of roadside IEDs. The LAVs proved useful in this role, one of their spectacular achievements being the escort of a massive "Mother of all [electrical] Generators" from the Jordanian border to Mosul without incident.[5]

Other missions were not so placid, and it did not take Polidoro long to receive his welcome message. "March 17, 20005, my first time out of the wire, I got blown up. . . ."

In the new operational area the Marines were spread thin. Second LAR, with an artillery battery serving as infantry, was based at Camp Korean Village. This force was responsible for securing some 10,000 square miles of nearly empty country with 387 miles of road, a handful of towns with populations up to 35,000 residents, and two major border crossings at Walid

(Syria) and Trebil (Jordan). Trebil was the crossing point for the critical Main Supply Route MOBILE, a six-lane highway along which three convoys per day brought fuel and other supplies. There were just too few men to adequately patrol—let alone secure—such a vast region. Polidoro: "At first, 90 percent of what we focused on was trying to make it a better place for the people. Around the third or fourth month, though, we were told to knock it off, so we backed off on that towards the end." Thereafter the battalion's efforts concentrated on keeping the main roads reasonably secure.[6]

Patrols or convoy escort in the LAVs was better because "The LAV is a little bit more spacious than a tank, and you can get away with a lot more 'cause there's storage in the area, but it is difficult to keep yourself clean. . . . Anything with fluids, you get all over you all the time. You get that, and you get rot all over you. Just the nature of the beast. . . . When you're rolling at 115 degrees, it's real tough." Sanitation consisted of a 40mm ammunition can with a plastic trash bag as a disposable liner.

Polidoro also found that the LAV was not as well designed for basic crew safety as the tanks. The turret basket is not self-contained as in a tank, and ". . . there's been multiple times I almost lost a foot in this thing, because where it's comfortable to me—behind the seat—there's no guard there."

On the patrols, some of which lasted sixteen hours, "Finding a position where I could sit or stand and see, but stay low enough to protect myself was damned near impossible. The seat—if I sit on it I'm too low, if I stand up I'm too high. What I would do is there's a little shelf—and by shelf I mean four inches of overhang—behind the seat of the vehicle commander. I'd wedge butt into that thing, as much as I could, and then there's sights right above the position. I'd put a water bottle next to that sight, and that'd be my chin rest. I'd rest my chin on a water bottle and have my M4 (carbine) up on top of the turret, with my finger on that thing with the seven-six-two [machine gun] ahead of that, with my hand on the water bottle looking UNDER the machine gun, so if something happened I'd get up and man the gun."

Polidoro thus found himself in the classic senior officer's dilemma; was his job to fire a weapon or direct the fight? "The M4 was for closed-bolt shots as your first escalation of force if you've got a vehicle or something coming for you, you use a closed-bolt weapon—one shot at a time. If that didn't work, you were supposed to move up to the machine gun, which was your open-bolt weapon, to stop and warn that car. And then the 25mm of course was for direct contact. . . . That was just too much for what we were doing. We were taking buildings down with the 25mm chain gun. . . ."

Korean Village was not supported by the usual civilian contractors, but operated by Marines and Navy Seabees, with Albanians contracted to provide sanitation services. The border posts were even more isolated.

"It wasn't as bad, though, as it was at the border sites, because we had Marines responsible for Walid and Trebil. Those were serviced by our logistics units, our battalion trains if you will. Once a week we'd send them out there to take care of those guys."

At the isolated posts "We got rocketed quite regularly. I wouldn't say effectively." The attacks inflicted only one KIA, but were a constant risk. To avoid a response, the insurgents would trigger the rockets with washing machine timers, and "They'd point it in the general direction of the camp. . . ."

Counter-battery fire, the usual reaction to such attacks, had to be used sparingly. Polidoro explained that "We only did it twice that I can think of in two years, because the collateral damage concerns were significant. Mostly they came out of neighborhoods. . . . Shot a lot of illum[ination] was really the only thing. Our primary response was patrol. Patrol, patrol, patrol, patrol, all the time."

Western al-Anbar Province remained an insurgent stronghold, and Operation MATADOR (8–19 May, 2005) was to be the first of several designed to gradually clear and secure the region where the rat lines crossed from Syria. It was also typical of the combined arms operations in the sparsely populated western region.

On 7 April insurgents, thought to be directed by al-Zarqawi, seized control of al-Qa'im, a city just east of the Syrian border, driving out Iraqi soldiers and police. From there they launched attacks—including firing missiles—at Camp Gannon, located in an abandoned warehouse complex in nearby Husaybah. Though the region north of the Euphrates was known to be largely dominated by *al-Qaedah*, the town of Ubaydi on the more accessible south bank of the river was thought to be relatively secure. The Marines of RCT 2 would secure the region around Ubaydi, cross the river on a bridge constructed by the Army's 814th Bridge Company, and scour the north bank for insurgents.

Company B, 2nd LAR and tanks quickly seized control of the old Ramana Bridge site (the bridge had been destroyed to hamper insurgent traffic) to cover the Army bridging operation, but the Marines quickly came under unexpectedly heavy attack. This triggered a fierce battle for "New Ubaydi," a town with paved streets and one- to three-story concrete houses, built to house workers at a nearby phosphate plant. While tanks assisted the infantry in clearing New Ubaydi, other infantry crossed the river in AAVs to begin scouring the north shore; because of accidents and other delays the ribbon bridge did not become operational until late in the day on 9 April.

Mines were as always the primary threat to tanks, and when a tank was disabled by a mine on 9 April it triggered heavy fighting. A recovery convoy was dispatched from the al-Qa'im base, but 7 km to the east plowed into an elaborate ambush. One Humvee was hit by a roadside bomb, and insurgents attacked with RPGs, small arms, and two suicide bomb vehicles. The Marines fought clear of the kill zone, but the M88 sent to recover the original tank was itself disabled by a mine. A tank towed the M88 back to base, and the disabled tank was later recovered.[7]

On the north shore the Marines swept to the west, clearing caves and locating weapon caches. On 14 May the Marines withdrew to the south bank, ending the operation. In MATADOR the insurgents displayed atypical organizational and tactical skills, and some were equipped with uniforms and body armor. As is the unfortunate case with many counterinsurgency operations, the overstretched Marines had no way to secure the cleared area, and insurgents quickly filtered back into the void.[8]

MATADOR was immediately followed by SQUEEZE PLAY in the environs of Baghdad, and then on 24–30 May Haditha was subjected to NEW MARKET, a sweep by infantry supported by tanks, AAVs, and LAVs.

More operations, often transitioning to "cordon and knock," followed in quick succession as the Marines strove to gain control of the Euphrates Valley and its rat lines. In a typical operation tanks and LAVs would sweep in to cordon a town or area, and infantry would sweep for insurgents. Many, such as CLEAR DECISION (30 April–5 May), were intended simply as clearing operations, but following Operation SWORD (Saif, 28 June–6 July), Hit became the first town to be permanently garrisoned.

A nagging problem, as in Vietnam, was the manpower consumed by maintaining and protecting logistical routes. The Marines had to secure hundreds of miles of roads through hostile territory, and the LAVs of 2nd

LAR were tasked with protecting both convoys and the Army engineers in their never-ending road maintenance.

Cooperation between Marines and the Army had become far more commonplace than in decades past, but policy still interfered. An Army infantry battalion was attached to the logistical command for escort and security. Polidoro:

> They were sick of doing convoy security all the time. We said "We're going to do a raid. We've got some great intel. We need some more people. Do you guys want to come and play?" They really wanted to come with us. We asked, "Do you mind if we take these infantrymen who know how to do this with us on our raid?" The answer was "No. They're Army guys." It wasn't their mission or role, so they wouldn't let them do it.[9]

CLEAR DECISION (30 April–5 May, 2005) was a typical example of a "cordon and knock" operation. Tanks and AAVs moved into position surrounding al-Karmah at 0300hours, followed by CH-46E helicopters dropping leaflets advising the civilian population what to expect and how to cooperate. At 0530 3/8 began the door-to-door search, while elements of 3rd Reconnaissance Battalion scoured the surrounding countryside. On 13 May the town was turned over to Iraqi government forces.

The Syrian border continued to be a trouble spot, and elements of 1st Tank Battalion, 2nd LAR, and 2nd and 4th AAV Battalions participated in Operation SPEAR, clearing the town of Karabilah.

RCT 8 was responsible for a large swath of country north of Fallujah, and operational units built around elements of the mechanized units, Team BRAWLER from B Company, 2nd Tank and Team GATOR from B Company, 2nd AAV Battalion conducted operations over the region. The two teams supported 1/8 in Operation KHANJAR (Dagger) in the southern Lake Tharthar (Buhayrat ath Tharthar) region, uncovering weapons caches and training facilities.[10]

The Coalition forces might to some extent control the operational pace and setting by launching operations designed to clear the Euphrates Valley and prevent insurgents from taking advantage of the impending force rotation, but the insurgents were by no means cowed. Near Haditha 3/25 had been engaged in fighting enemy units using mortars and small arms. On 1 August insurgents struck two scout-sniper teams from 3/25 near Barwanah, northwest of Haditha on the left bank of the Euphrates. They quickly overwhelmed the teams, killing five and carrying away a sixth man whose body was found the following day.

Forces allocated for Operation LIGHTNING STRIKE II were quickly reallocated to QUICK (3–11 August, 2005), a large cordon and search centered on the villages of Haqlaniyah (where the body of the abducted Marine was found) and Barwanah. In a common pattern 2nd LAR established a loose cordon around the area, infantry and tank units moved in, and Force Recon provided raiding parties and sniper support. The heaviest loss occurred on the first day of the operation, when a roadside IED struck an AAV carrying Marines from L/3/25. The huge explosion blew the AAV onto its side, killing fifteen crew and passengers.

On 4 August two companies from 3/2, supported by 2nd Platoon, A/1st Tank moved north through Barwanah on the north bank of the river, while two companies from 3/25, supported by 3rd Platoon and the Headquarters Section of the tank company moved in parallel through Haqlaniyah on the south bank. The Marines encountered only light resistance, but the heavy loss of life–particularly among a Reserve unit–attracted considerable media attention.

By now the fighting had settled into the seemingly interminable series of clearing operations that characterize any counterinsurgency campaign. The task was made particularly complex by standing up the new Iraqi military and police units, Operation GUARDIAN SHIELD (a blanket operation to cover another force rotation), and Operation LIBERTY EXPRESS (protection of the polling places for a constitutional referendum on 15 October and national elections on 15 December). Sunni moderate factions had "sat out" the prior elections, but this protest had backfired, conceding control of the central government to Shi'ite and Kurdish factions. In this election cycle the fear was that extremists would once again discourage Sunni participation, creating a permanent split among the various factions vying for control of the new Iraq.

The Marines estimated that despite diverse commitments, they could retain the initiative and disrupt insurgent efforts. The Multi-National Force blanket operation, SAYAID II, would establish indigenous forces in al-

Anbar, secure the region along the Syrian border, and interdict insurgent activities all along the Euphrates River valley. Many local operations would exploit sweeps by establishing permanent presences in an increasing number of towns, and civil affairs programs would concentrate upon encouraging anti-insurgent sentiment among the Sunni leaders and populace. Forces were still inadequate to either control the Syrian border or to adequately control the large region along the river valley. Part of the effort to restrict insurgent traffic was to destroy selected bridges across the Euphrates with air strikes and Army MLRS rocket attacks, and to establish random checkpoints along major roads on the south bank.

Thus far the Marines had concentrated on securing the more densely populated Euphrates Valley and the crossings into Syria. The huge, empty expanses of desert south of the river bordered by southern Syria, Jordan, and Saudi Arabia is traversed by only one major highway, Route 10, which crosses from Jordan near Tribil, passes through ar-Rutbah, and joins Route 12—the main road that parallels the river valley—near ar-Ramadi. The logical units to patrol and secure this lengthy and desolate highway were the LAR battalions, but they had been committed to almost constant action along the highways to the north and securing logistical routes.

The only consistent American presence was at Camp Korean Village, or Camp KV, a desolate outpost near ar-Rutbah, the nexus of most of the secondary roads with Route 10.[11] Camp KV was regularly resupplied by convoys, but living conditions for the troops stationed there remained bleak.

In the pre-dawn hours of 11 September 2nd LAR established a cordon around ar-Rutbah, the beginning of Operation CYCLONE (Zoba'a). Infantry forces swept through the town from two directions, the north (Marine Force Recon and Iraqi Special Forces) and the south (K/3/6 supported by AAVs and a platoon from C Company, 2nd LAR), targeting pre-selected houses. This was to be 2nd LAR's final significant operation before rotating out in September.

By October reliable Iraqi units were arriving in some strength, to secure the Syrian border and to garrison towns beginning with Hit, Hadithah, and Rawah. To draw insurgent attention from this movement and preparations for major operations yet to come, between 1 and 8 October the newly-arrived 3/6 executed Operation IRON FIST (Kabda bil Hadid), a sweep of

Sadah, a town of 20,000 persons about 12km from the Syrian border, and the eastern sector of nearby Karabilah.

On 2 October three rifle companies from 1/3 moved through the town of Sadah, supported by platoons of tanks and AAVs; resistance was atypically direct, with insurgents attempting counterattacks with small arms and RPG and vehicle-mounted squads, as well as the usual roadside bombs. In at least one case tank fire was used to destroy a car bomb.[12] Considerable propaganda always resulted from civilian injuries, as in one case where insurgents forced their way into a building still occupied by civilians. Probably unaware of the non-combatants, the Marines used a round from an M1 tank to demolish the building, wounding five civilians.[13] After securing Sadah, the Marines and supporting soldiers blocked roads connecting the two towns, and set up defensive positions in a wadi between the two.

The fighting proved much heavier in Karabilah, where tanks and air strikes were used to blast insurgent positions.[14] Again the insurgents along the Syrian border were heavily armed, with mortars and other heavy weapons, and more proficient than their brethren around Baghdad. One NCO, a Sergeant Lybarger, remarked, "They were as good as our guys were. I wanted to kill them before they teach all the other guys how to do that."[15] Three days were required to reduce Karabilah, and sporadic fighting by patrols continued for weeks while a permanent base was constructed.

Even while this operation was underway, the Marines returned to another chronic sore spot. The large Operation RIVER GATE (Bawwabatu Annaher, October 2005) centered on Haditha, Haqlaniyah, and Barwanah, and was designed to pave the way for a permanent presence in this troublesome region. In a large and complex operation Iraqi special forces units provided a broad screen. The Army's 3rd Battalion, 504th Parachute Infantry, transported by Marine helicopters and supported by 3rd Platoon, B Company, 1st Tank and Iraqi units occupied Haqlaniyah. The 1st Platoon and Headquarters of the tank company supported 3/1 and Iraqi units in the occupation of Hadithah. 1st LAR occupied Barwanah.

This area proved to be part of an insurgent base area, with not only the usual bomb factories, but a large propaganda mill equipped to produce videotapes and CDs in quantity. After the operation Marine engineers moved in to construct patrol bases in all three towns.

The Marines provided both protective and limited logistical support for the constitutional referendum, and took preemptive measures such as placing indirect fire on likely launch points for mortar or rocket attacks.

As more Iraqi units became available they were increasingly incorporated into operations, proving particularly useful in identifying foreign insurgents who were otherwise invisible to American troops. In Operation STEEL CURTAIN (al-Hajip Elfulathi) a combined force of Marine Corps (including 1st LAR), Army, and Iraqi troops swept from west to east through Husaybah between 1 and 3 November. The task force continued east through Western Karabalah, then on to again clear Ubaydi. This time a permanent presence was established in the cleared towns.

With the completion of LIBERTY EXPRESS, in the New Year the mission shifted to outright suppression of the insurgency. In January through April of 2006 I MEF relieved II MEF, covered by a number of small operations designed to prevent insurgents from taking advantage of the relief. 1st LAR assumed responsibility for the ar-Rutbah region, and from 18 to 25 January prosecuted WESTERN SHIELD to clear and secure the town. As part of the process a two-meter high earthen berm was constructed around the town, with only three traffic control points to provide access to the city, thus denying it to the insurgency as a logistical base and transit point.

As the heaviest fighting wound down, an increasing number of Marine Corps units found themselves serving as provisional military police. Artillery units in particular were diverted, starting in 2004 with the entire 3/11 howitzer battalion. Mummey's nephew did two tours. "He was a gunbunny, but he never pulled a lanyard. He was always doing convoy security, or working with an Iraqi police team. He said 'Uncle Mike, it's hard. You come back in and all of a sudden you gotta relax. You got the internet, you got your little iPod ™ you listen to your music, and go over to the phone center and make a phone call. Then all of a sudden you go to the mission brief, and you're gearing up, checking your gun, making sure you got the call-signs, got your medical kit fit to go, and then you jump on this convoy away to get blown up, you know."

Line and headquarters companies from both 4th Tank and 4th AAV Reserve units would serve as provisional MPs, while small boat detachments for dam security would be provided by AAV and LAR units. The other shortage was in civil affairs specialists. The two existing units were grossly overburdened by continuous rotations, so two provisional groups were formed by retraining other specialists. The insurgency also shifted tactics, relying increasingly upon IEDs to disrupt civil affairs efforts and attack individual vehicles and convoys, increasing the demand for engineer and EOD specialists.

Gunnery Sergeant Michael Martinez of Bravo Company, 4th Tank was uniquely qualified for MP duty. Without the inclination or money to attend college, he decided as a high-school freshman to join the Marines, and bided his time until old enough to enlist. After active duty in the infantry, he became a US Marshall and a Reserve MP, and then went to an LAV unit. In 1998 the Marshall Service transferred him to Washington State, so he became a tank commander.

Following notification in mid-2004, the tank company, with a platoon of Reserve MPs, was formed into a Provisional MP Company. Martinez said of he and another Gunny who was in civilian law enforcement, "Between us we came up with a training schedule. In October through December did a whole other's years worth of drills. . . ." Martinez included both military "lessons learned" materials and training from his civilian job and "We put together a bunch of classes on clearing buildings, clearing rooms. Some basic stuff on patrolling packages, driving, a little convoy stuff." Drills included reaction to IEDs and small arms attacks, and casualty evacuation. In contrast, language and cultural training was "was pretty rudimentary stuff. It certainly wasn't in depth."

The US military was still feeling its way into the strange environment, and trying to incorporate models from other occupation experiences. The tankers trained in British-style satellite patrolling on foot. "Instead of just getting into a column-of-twos and walking through town, it's sort of a decentralized patrol is the best way to put it. You're flowing in and around objects instead of staying one behind the other walking like ducks in a row. . . . You try not to put everybody on the same path, so that you reduce vulnerability."

Martinez's Provisional MP Company was based outside Fallujah. Living conditions were good, with most men housed in a metal building with indoor plumbing and air-conditioning that "mostly worked." "Chow was way too plentiful. It wasn't overly good, but there was a lot of it."

The company's primary mission was to escort convoys over a vast region. "We also did prisoner releases from Abu Ghraib up to Ramadi pretty frequently. Those were daytime missions, those and escort missions for inductees into the Iraqi National Guard."

The first incident was near the prison. "It was in a place that was heavily IED'ed. . . . This one turned out to be a dud. The kicker charge went off, and it was enough to throw the one-five-five rounds up on the tarmac. We stopped and cordoned it, and nobody was hurt. . . ."[16]

On the convoys, "Usually we were escorting TCN vehicles—Third Country National drivers—logistical convoys. There was a marshalling yard pretty close to the building we were sleeping in. We would go over there and meet with the guy that ran that. He would tell us how many vehicles we were taking." The convoys moved at night since the curfew allowed the convoys to avoid civilian traffic. "And at that time the rules, keeping cars away and running cars off the road, that kind of stuff, was still heavily in effect, and they just didn't want us running those roads during the day and impacting the locals."

The nighttime runs minimized attacks on the convoys, since the insurgents were disorganized and most attacks occurred during the day. "They popped IEDs on us a couple of times. None of the small arms fire was ever really directed at us. We were just kind of in the area. The way we ran was blacked out, and then the semis (contractor cargo trucks) had lights."

The usual configuration was for a two-vehicle lead element to proceed about two kilometers ahead of the main body. "We got hit with IEDs on four occasions. They always hit our lead vehicle."

One day the escort became separated from the convoy, a resupply run for a Jordanian field hospital with doctors and nurses aboard, in the Korean Village—ar-Rutbah region. "I told everybody to stop so we could get everybody back together, and they stopped right adjacent to that fifty-five gallon drum, and it blew up. Center-punched the vehicle. . . ." The explosion flattened the tires and tore away coolant hoses and other engine parts.

"We had one Marine that got wounded, took a piece of shrapnel to his forehead. . . . That was the first MAK vehicle our unit had. . . . Everybody inside was fine. The kid in the gunner's turret caught a piece of shrapnel. . . . I think they said it was three one-hundred-twenty-two millimeter rockets that were bundled, and they were pretty much right next to it when they went off. It was in a fifty-five gallon [drum] by the side of the road."

Near the Jordanian border "It was pretty desolate out there. . . . They had just passed through a little settlement, it wasn't really even enough to be a village, just a couple of houses. The trigger man was in one of those houses."

Martinez, in command of the convoy since the Platoon Leader was away at a briefing, was riding in a truck just behind the lead element. The usual drill was to keep rolling, but the immobilized vehicle and a casualty halted the convoy. "I rolled up there with my seven-ton, we attached the seven-ton onto the Humvee, dragged it about four-hundred yards down the road just in case they had any daisy-chains waiting to hit the rest. We got down there, we assessed the wounded, got him on the road with the corpsman. Then we changed the tires on that vehicle. By the time we got the tires changed I rolled up my vehicles from the rear. We set a perimeter and we started clearing everything immediately adjacent to where the explosion had happened.

"We caught some guys who were peeking out at us through some windows in some buildings that were a couple of hundred meters away. . . . We detained three, two adults and a juvenile, all males." The prisoners were questioned and released at Korean Village. The provisional MPs learned that 2nd LAR later captured or killed some of the actual triggermen.

By 2006 Iraqi forces had assumed much of the responsibilities at ar-Rutbah and Camp KV, gradually replacing the Marines of 2nd and 3rd LAR.

Major John Polidoro returned for a second tour in August 2006, still as XO of 2nd LAR. He felt that this time the unit was far better prepared, with better Arabic language and cultural awareness training, and three weeks training in urban combat as part of the MOJAVE VIPER program. Upon arrival, Polidoro found that the battalion was to be split between two separate Areas of Operation. The bulk of the battalion—two LAR companies, most of the H&S Company, an attached infantry company, and a section of artillery—went to Rawah in the Euphrates valley. Polidoro took one company, a company of MPs, and part of the H&S Company back to Korean Village. Again, Polidoro found himself in an "economy of force" mission, trying to control border crossings. Too few men meant that he could set up traffic control points around the bermed town of ar-Rutbah, but could not adequately patrol the town.[17]

Living conditions had actually deteriorated at Korean Village. Captain Jason Gaddy had served as the battalion H&S Company CO on the prior tour, and had encountered problems with contractor services. "It was worse than the first deployment, as far as logistics. Marines were still living in tents. There was no real electrical grid. The kitchen was bare bones. The cooks had two burners and a griddle and had to cook for 900 people. . . . I don't know if it was because when the Strykers pulled out they took all of their stuff. Third LAR came up and they were only there for like a month."[18] Gaddy faced a continuing uphill battle with logistics, dealing with the Army, contractors, and Navy Seabees.

Despite the limitations of manpower, the force at Korean Village was able to focus more on interacting with the locals and exploiting intelligence, rather than road security. Cooperation with the Army infantry battalion in the region, and a local militia group, was far better, but the mission was still just to "keep the lid on." Polidoro could observe insurgent activity in the town, but ". . . just because we saw something in town didn't mean we were going to run in there willy-nilly."[19]

Still, ". . . it was ugly, very, very ugly," said Polidoro. "That was complex ambushes, complex IED chains. . . . and there was a lot of fighting going on." In November "we had about thirty-five or forty insurgents open up on a checkpoint into the town of Riphah. We had a Marine shot in the neck in the opening stages of that one, and we had to move some significant forces down there to handle them. . . ."

IEDs remained the weapon of choice for the insurgents, and for the patrols "Routine usually meant they'd go out and somebody would get arrested or they'd get in a firefight or blown up. A day didn't go by when there wasn't some type of activity. . . . I would say the average incident rate in 2005 was about three events a day, which was either an IED find, an IED explosion, or . . . (we) detained somebody. In 2006–2007 we averaged close to six."

Suicide vehicles remained a threat, and the problem of defending against them was exacerbated by the appalling driving habits of the Iraqis, who blundered into roadblocks and vehicles. "Nine times out of ten this wasn't intentional. The Marines were convinced that the Iraqis were drunk all the time, or they're blind." The use of pen flares, pocket sized devices that could be fired at inattentive drivers was one solution. Polidoro recalled only three incidents of actual insurgent attacks using vehicles, and one was a vehicle caught at a roadblock.[20] Usually the Marines would simply move off the road to avoid potential threats.

Despite the improvements Polidoro concluded that if struck by an IED he would rather be in an up-armored Humvee than an LAV. "They're not that thick. The one thing you've got in your favor is they're very high off the ground. When you're up high it's not too, too bad. It's basically like being in the Liberty Bell, and you hit it with a hammer really, really hard. It really shakes you pretty well. Significantly loud. I lost a lot of hearing because of these things.

"The first response is 'What the hell just happened?' The second response is 'Is anybody hurt?' The third response is 'I'm gonna kill the guy that just did that!'"

That effort was usually fruitless. "We captured a bunch of guys, but you never can prove exactly who they were. You gotta find the smokin' gun."

Casualties were usually minor shrapnel wounds. "The biggest problem with the LAV in an IED blast is that tires are solid rubber. So they catch on fire. When the tires catch on fire, you can't get them out. So the thing burns to the ground, and it becomes a low-rider pretty quick. And then it's a catastrophic loss of the vehicle."

Gaddy's logisticians continued to improvise:

The COP was in such close proximity to Rawah that we pushed hot chow out to the Marines out there too. I worked my cooks pretty hard and they didn't have a lot of equipment. One of my favorite things was when the Army lieutenant came for Thanksgiving and said, "Hey, sir. I got us 60 Turkeys for Thanksgiving." I was like, "60 whole turkeys?" He said, "Yeah." He was excited. He was a green lieutenant and I had no patience. I was like, "60 birds. That's great. How are you going to cook them? How do you think we're going to cook them?" He just looked at me dumbfounded and said, "Sir. I got us 60 turkeys." I was like, "Have you seen an oven? Even if I had an oven I would have had to start last week. I don't have turkey fryers. . . .

I was getting ready to send the crap back and say, "Thanks, but no thanks." The battalion maintenance chief, Master Gunnery Sergeant Tooth (sp?), was like, "Hey sir. Don't get rid of them. We'll figure out a way to cook them." I sat down with my chief and started going through ways of how we could cook turkeys. I was like,

"Okay. We're not going to get an oven. We're not going to get turkey fryers." Finally, I remembered, "Hey. I've smoked a bird." I called down to my battalion maintenance officer and said, "Hey Mike. Send me Staff Sergeant Delany." Staff Sergeant Delany was the welding chief. He came up and I said, "Hey man. I need a hero." He said, "What do you need?" I said, "I need a smoker. Can you build a smoker?" We drew it out on paper, I told him what I needed and three days later the tram, which is a 10,000 pound forklift, brings up this monster of a smoker. . . .

We fit them all in there and about 30 hams as well. We used that thing quite frequently. It had the fire boxes down below and it was vented. It was absolutely awesome. It was about seven-feet tall by about 10-feet long and three-feet deep. It had four trays. It was awesome. The Marines got turkey. [21]

In December most of 2nd LAR (less one company that stayed in Korean Village) was reunited in Rawah. By this time the slow and often painful process of building an Iraqi force was beginning to pay dividends, with development of an aggressive Iraqi Army battalion commander, and a hundred-fold increase in the number of local policemen. Nevertheless, the local politics could be Byzantine. Marines of the LAR line companies were quartered with the local Sunni police and thought them quite professional, but the Shi'a Army battalion that had occupied the area was prone to prisoner abuse. A new battalion was more professional, but mutual trust with the civilians was slow to recover.[22]

Harassment continued, and "We were rocketed quite regularly," said Polidoro, "usually about five o'clock at night, usually by about the same guy, which was really disruptive of our chow pattern. Kind of pissed us off a little bit." Counter-battery radar has difficulty in detecting low-trajectory projectiles, so the response was limited.

The Marines established a permanent presence with traffic control points and bases inside Rawah, and patrolled aggressively. Intelligence, always a prime factor in counterinsurgency operations, was beefed up and pushed down to the company level, with good results. Polidoro: "On one night alone we rolled up 27 guys. We had a firefight that day and captured a guy who was a messenger for one of the head insurgents in town. He gave up a lot of information, so we acted on that and were able to snatch up all those guys. The indication we had that that worked out was that before we snatched up all those guys, there were death threats out there against women in Anah that said if they didn't wear veils and all that other stuff, they would be shot. Well, we rolled all these guys up and the next day all the girls were in their jeans and they had makeup on." Such was the peculiar measure of success.[23]

The downside of the new operational procedures was that wounds tended to be more severe since much of the patrolling was on foot, and the men were vulnerable to snipers and IEDs. Medical corpsmen were a much more pervasive presence than ever before. "We would put one corpsman out per platoon" said Polidoro. "We tried to roll five per company, which would be Headquarters Platoon plus the three [line platoons] plus a company corpsman. We had a battalion aid station run by a Navy lieutenant. . . . The second [battalion surgeon] was fantastic, a guy named Drew Bailey. He was phenomenally good. He put thirty-seven pints of blood through one kid; kept him alive." Lt. Andrew Kinard "Stepped on two one-five-fives [rounds], which took him apart. He should have died on the table, no question about it."

In early 2007 the Marines finally withdrew, but they would soon return. Third LAR would return to this stomping ground in October of the same year. For years the LAR battalions would rotate in and out of Iraq.

Promoted to Gunnery Sergeant, Sam Crabtree had assumed the senior role of Tank Leader for Delta Company, 4th Tank. "The tank leader always knows what's going on. He's in charge of making sure everything happens. . . .Beans, bullets and band-aids." Duties are all-encompassing, from coordinating communications to delivery of Logistical Packages—standardized issues of fuel, food, water, ammo, and everything else the formation needs.

"Being the Tank Leader, I got to go over with the CO and the XO for a PDSS (Pre-Deployment Site Survey) to see what was going on in Iraq before we deployed." This included tours of the future operational area, and conferences with the personnel they would replace. "What are you guys really doing? . . . What did you think was going to happen and you were totally wrong? What can we do so we don't go through the same problems?" The main body of the company deployed in October 2007.

The three tank platoons of Delta Company were immediately scattered in widely separate areas of al-Anbar Province. "We had one platoon in Rah, one platoon in Haditha, and we had another platoon in Hit. Each platoon of tanks was chopped (placed under the operational control of another command) to that area, to whoever was in charge, which was generally a grunt battalion. They would do whatever they needed to do." The platoons would provide Quick Reaction Strike forces, cover sniper teams, and provide convoy security.

"Headquarters, we would do what was called Team Tank. . . . Regiment would pick an area, and we would take ninety per cent of Headquarters with our seven-tons, Humvees, HETTs from the Army, and search them for weapons or just ammo. Anything that we could do." In the empty desert "Sometimes we'd go for two or three days and wouldn't see any structures, just roads, palm groves, or something." "If we saw vics, we'd roll 'em up, snap 'em up, check 'em out." Interpreters were attached, and personnel had been trained in interrogation techniques.

"It was called 'Everybody's a collector.' We'd ask them, 'Have you seen any of the bad guys? Where are you going? How are prices? Has the price for water gone up?' Just general questions, to try and get a feel for what's goin' in the area. 'Where'd you come from and where you goin'?'" The team might spend weeks checking past insurgent weapon cache sites, and the local caves for evidence of activity.

After weeks of such activity "Even when you get back to the FOB, yeah, it's nice. You get good chow, get cleaned up, but at the end of the day you're still in Iraq. You sleep, but you don't sleep."

To keep contact with the isolated platoons, "Me and the CO would try to visit the [platoons] every other month. . . . We'd go there, see how they're goin', any problems they have. Also to get a feel for how's their equipment holding up. . . . How's hygiene? How's morale?" Conditions were different at each position. "Down at Hit they would go out all day long, but come back all the time. At Haditha . . . they were covering the sniper teams while they were trying to get into position or out. Sometimes they would sit there for two days. . . . Just scan the desert for hours at a time."

One relief from the sameness was the assignment to protect the annual Shi'ite *haj*, or pilgrimage. "We had to go down to some FOB and we'd provide security not on the roads, but out away from the roads."

This tour coincided with the introduction of the MRAPs, Mine-Resistant Ambush-Protected armored trucks. In addition, "We also had a roller that mounted on the front of a seven-ton that we'd put in front of the convoy." Various vehicles—mostly tanks—were struck by IEDs that "Beat some guys around. But no serious injuries." To counteract the new command-detonated mines triggered by cell phones "We had jammers on our vehicles." Other age-old precautions were still effective: varying routes and routines, avoiding likely ambush sites.

Crabtree ended up operating one of the new MRAPs as more were issued to the regiment. "The intent was to phase out the Humvees. It's comfortable when it's not moving. It's got a lot of cool stuff, (stowage) boxes and everything. The trouble is you go off road in those versions of the MRAPs, it beats the shit out of you. They put the seven-ton suspension on the MRAPs. If you're going ten miles an hour off-road, you're hauling ass. I didn't want an MRAP. I wanted to stay with my Humvee because it was more maneuverable, and it could hold up to the speeds.

"[The MRAP] was awesome to live in, because it had AC that worked really good. . . . The nice thing about the MRAP too is you're higher up, so your gunner up top he can see more of what's going on. More situational awareness than a Humvee turret.

"But every op we went on, we broke one. The suspension broke real easy. It wasn't designed to hold up to what we were doing."

Crabtree thought the seven-ton was the best vehicle. "It rarely broke down. It had good suspension, and of course it's just a flat bed. You can configure it like you want. It's kind of tall, so high-visibility. With high-visibility you're more of a target, but if I can see you sooner, I can kill you sooner.

"When you're all buttoned-up in the MRAP the comm system didn't work very well. It was hard to hear, and you got all that armor around you, you can't see as much. . . . I had trouble seeing through the bullet-proof glass with NVGs [Night Vision Goggles] on."

Routine COIN operations like convoy escort would consume progressively greater portions of the tanker's efforts until the end of the deployment in March 2008. As the Marine presence wound down, the tank units were increasingly deployed in such roles as Provisional MPs and many other roles. Because that's what Marines do.

Another officer new to Iraq was Marc Tucker. He had wrestled in college, but damaged his elbow, and "I just finally came to the conclusion that I was in school to wrestle versus being in school to be in school." He worked for half a year, and then joined the Marines. "I knew that being a Marine was what I wanted to do, and that was it. . . . I knew that I wanted to be an infantryman, and an infantryman only. . . ." After nearly four years in the infantry, he was selected to be an instructor at the Officer's Basic School, and three years later signed up for the Marine Corps Enlisted Commissioning Education Program (MECEP) and graduated from college. By 2007 he was back at Basic School, this time as a candidate. Tucker's performance was sufficient to win him the single tank school slot upon graduation, over 100 or so other applicants.

After tank school he became a very junior staff officer in 1st Tank in early 2008, since his assigned company was in Iraq. Tucker was a bit uneasy; though experienced, he was "uncertain what the future held" in armor. "It was exciting and nerve-wracking at the same time" since the expectations for former enlisted men were so high. After a few months he was "offered the opportunity" to go to Iraq as part of a Military Transition Team. After months of training at the last minute he was instead hastily trained by the US Border Patrol, and assigned to a Port of Entry Team, training Iraqis to run one of the major logistical entry points from Jordan. Most of the original personnel were from 1st Tank. "We were handed a bunch of references and told to go forth and prosper." Major problems included smugglers bringing in copper used as a liner in shaped charges, and initiation devices for IEDs. "We had to figure out very quickly that there was a certain acceptable level of corruption. . . . What is morally and ethically correct inside the United States doesn't necessarily play the same way in Iraq."

Any romantic notions of battles with smugglers soon were replaced by the tedium of routine border patrols in Humvees, and training Iraqis. On border patrols, "It could take us a day to get to the farthest outpost. . . . We'd spend two or three days hitting all the specific posts." In the open desert it was simple for the smugglers to evade the patrols. "Just as we started figuring it out it was time for us to leave, time for a fresh crop of bright-eyed bushy-tailed, (an) unjaded team, to head in and spend the next few months trying to figure things out."

NOTES

1 Fike, *Interview with MAJ Robert Bodisch*, p. 8

2 Estes, *US Marine Corps Operations in Iraq, 2003–2005*, p.110.

3 Lawrence Lessard, *Interview With Major John Polidoro*, Part I, p. 3.

4 Ibid, p. 4.

5 Estes, *US Marine Corps Operations in Iraq, 2003–2005*, p. 115.

6 Lessard, *Interview With Major John Polidoro*, Part I, p. 5–7.

7 http://mypetjawa.mu.nu/archives/081669.php

8 http://en.wikipedia.org/wiki/Battle_of_Al_Qaim

9 Lessard, *Interview With Major John Polidoro*, Part I, p. 10.

10 Note that this operation name was also used in Afghanistan in 2009.

11 The camp reportedly housed Korean laborers who constructed the highway during Saddam Hussein's reign. The facility was established as FOB BUZZ by Army cavalrymen who manned it from the invasion until they were relieved by the Marines. No one knows who coined the KV name.

12 http://www.defenselink.mil/news/newsarticle.aspx?id=18174

13 http://www.globalsecurity.org/military/ops/oif-iron-fist_2005.htm

14 http://newsblaze.com/story/20051007094130nnnn.nb/topstory.html

15 http://www.longwarjournal.org/archives/2005/10/operation_iron.php#ixzz0l3vIXeNk&c

16 Probably actually old Iraqi Army 152mm rounds.

17 Lessard, *Interview With Major John Polidoro*, Part II, p. 4–6.

18 Fike, *Interview with Major Jason Gaddy*, p.12.

19 Lessard, *Interview With Major John Polidoro*, Part II, p. 7.

20 Author, Polidoro interview

21 Fike, *Interview with Major Jason Gaddy*, p.13.

22 Lessard, *Interview With Major John Polidoro*, Part II, p. 9–10.

23 Ibid, p. 8.

CHAPTER TWELVE

Afghanistan—Winding Down

There is an opponent in a war, so one cannot progress just as one wishes.—Admiral Matome Ugaki, Imperial Japanese Navy

No war is over until the enemy says it's over. We may think it over, we may declare it over, but in fact, the enemy gets a vote.—General James Mattis, USMC

AS THE U.S. WAR EFFORT was redirected in 2009, President Obama authorized a controversial major troop build-up in Afghanistan. As part of the surge in troop strength, D Company, 2nd LAR and a headquarters detachment were assigned to 2nd MEB, in central Afghanistan. In early June conditions were typical. The battalion commander wrote to families:

> Hello from Central Afghanistan, where we are in the "Season of 120 Days of Wind." If you are wondering what this season means (no, it's not when Congress is in session!) for your Marine or Sailor, it is dust storms or 'brown outs' almost daily beginning at about 1430. The dirt is everywhere and in everything. It has to be considered in our operations as well. Another unfortunate impact is that our showers are not collocated with our tents, so it's not uncommon to see a Marine or Sailor returning from the shower dirtier than when he went as the dirt sticks to him on his return trip; picture a powdered donut.[1]

A first major action was Operation KHANJAR (Sword Strike), in southern Helmand Province. The avowed purpose was to secure the large area to allow for meaningful participation in the national elections of August 2009. The region was also a transit point for much of the opium that funds the Taliban, and an infiltration route for foreign fighters seeking to avoid Coalition troops on the border with northwest Pakistan. Infantry battalions were assigned to the more rugged northern part of the operating area, while LtCol Tom Grattan's 2nd LAR was assigned to the southern district. British troops and Danish tankers would coordinate their activities in surrounding areas.

On 9 July Marine infantry began moving on suspected Taliban strongholds in the province, exercising tactical restraint to avoid civilian casualties. The infantry battalions were met with varying degrees of resistance, from roadside bombs to ambushes. On 9 July 2009 a seventy-vehicle convoy from 2nd LAR arrived outside the village of Khanashin, the capital of the southern Rig District closest to the border with Pakistan's Baluchistan Province. The town had long been a Taliban stronghold, and Coalition forces had never succeeded in establishing a base nearby.

The battalion met with only one incident of sniper fire. The Marines chose to negotiate rather than force an entry; village leaders urged the Marines to occupy the town, but the locals stayed indoors, wary of possible fighting.[2]

The Marines settled in for months of patrolling, trying to interdict drug trafficking, and occasional IED and mortar attacks that inflicted deaths and injuries. On 10 July an IED detonated beside an LAV, killing MSgt Jerome D. Hatfield, the Delta Company operations chief, and crewman LCpl Pedro Barbozaflores. On 23 July mortar fire on the E Company compound killed Sgt Ryan Lane.[3]

For the Marines in Afghanistan, the mission remained one of winning the trust and acceptance of the people, who in this region of Afghanistan have not seen representatives of a real government in decades. On 29 July a large patrol moved toward another village, seeking a site to establish an Afghan Security Force post. Marines sought out village elders to negotiate a site—which would inevitably mean loss of farming land. In exchange, small gestures meant a great deal. First Lieutenant Joseph R. Gazman:

Even when we came into the town, one of the elders was concerned that our presence was scaring the children. The Marines were instructed to position themselves in less visible places in the town. The elders want to make sure it is a peaceful coexistence. Many of the people here have been detached from the presence of government for such a long time, so they haven't really seen a large [military] presence. For someone like that, this situation can be understandably intimidating.[4]

The period just prior to the 20 August elections saw the inevitable upsurge in Taliban violence, but the Marines preempted many of their efforts, preventing the enemy from moving mortars and rockets within firing range of polling sites. Attacks by 2nd LAR killed Taliban leaders, but they lost an EOD technician (GySgt Adam Benjamin) attached to Charlie Company to an IED.[5]

On 3 September Delta Company moved in to occupy a compound formerly occupied by the Taliban, and slated for use by Afghan security forces. A concealed IED detonated, killing LCpl Chris Baltazar and HM3 Ben Castiglione; LCpl Christopher Fowlkes later died his wounds. Corpsman HM2 Tommy Peterson and LCpl Christopher Gisbrecht were wounded in the blast.[6]

Benjamin P. Castiglione:

"The military life is not easy," Castiglione told a newspaper last winter, "but I believe that I am a stronger person for it. The hardships I have dealt with were worth what I have learned and the bonds I have with the Marines in my platoon. When my platoon and I have downtime and talk and mess around with each other, it's like one big hilarious, dysfunctional family — and it's a blast. We take care of each other."[7]

From the earliest days the US military resisted sending tanks into Afghanistan, though NATO forces deployed tanks in the country beginning in December 2006. There were two NATO tank units, a five-tank platoon from Denmark and a fifteen-tank squadron from the Canadian Lord Strathcona Horse (Royal Canadians) attached to 1st Battalion, Princess Patricia's Canadian Light Infantry and 1st Battalion, The Royal Canadian Regiment.

Danish tanks of the first battalion, *Jysdske Dragonregiment* (Jutland Dragoons Regiment) operated in support of British and Danish forces in Helmand Province, typically as two-tank sections. Both Danish and Canadian sources emphasized that contrary to the expectations of "experts," the local population supported the utilization of tanks. In a 5 January 2008 engagement with strong Taliban forces, three Danish tanks ". . . reduced the need for air support. Tank fire, which is frightenly (*sic*) accurate, pentetrates (*sic*) walls but usually does not level a mud-brick compound the way large bombs dropped by aircraft can. This makes reconstruction in the area far easier once the Taliban have been removed."[8]

In the Canadian experience, ". . . suggestions that the use of tanks has alienated the local populace more than other weapon systems have proven completely unfounded. Since commencing combat operations nine months ago, Canadian tanks have killed dozens of insurgents in battles throughout Kandahar Province, yet there has been no suggestion of civilian deaths attributed to tank fire during this entire period. Equipped with a fire control system that allows our soldiers to acquire and engage targets with precision and discrimination, by day and by night, the Leopard tank has in many instances reduced the requirement for aerial bombardment and indirect fire, which have proven to be blunt instruments." The analysis detailed how tanks could be used effectively not only in open terrain—where the Taliban's efforts to confront them proved suicidal—but in the populous areas.[9]

Tanks of course also prove far more resistant to the IEDs so favored as weapons by the Taliban. A Danish tank struck a large IED in July 2008, killing the driver and wounding three other crewmen. IED strikes on at least two other Danish tanks resulted in no casualties.

The Marine Corps originally deployed only two M88 retrievers to Afghanistan, as no recovery wheeled vehicle could tow the heavy MRAP (Mine Resistant Ambush Protected) trucks. The MRAP program came about through a cold, Defense Department calculation: ". . . the real dollar costs of care and replacement—adjusted for enlisted casualties average $500,000 while officers, depending upon their military occupation range from one to two million dollars each. This meant the average light tactical vehicle with one officer and four enlisted personnel was protecting 2.5 million dollars of the DOD's budget."[10]

The 2nd Combat Engineer Battalion deployed the new Assault Breacher

Vehicles (ABV), based on the M1 tank chassis. The ABV was developed by the Marine Corps following the cancellation of the Army's similar Grizzly engineer vehicle program. Based on a refurbished M1 tank chassis, the vehicle is equipped with plows, line charges, and is capable of remote-controlled operation; it is strictly an engineer asset. Ironically, this type of vehicle was first proposed by one of the Corps' farsighted armor advocates, Col. Arthur "Jeb" Stuart, in 1949.[11]

Another factor that deterred the use of tanks was transportation. Mike Mummey explained, "Word is the Army won't give up the [air]lift. You can only move one tank at a time on a C-17. There are gonna be some tankers go over there; they're gonna be on MRAPs."

In late 2010 Major General Richard P. Mills of Regional Command—Southwest requested tank support. Pentagon spokesman Col. Dave Lapan cited ". . . increased mobility, the increased firepower, [and] because of the optics the tanks have." He went on to explain that "Tanks are more accurate than artillery, mortars or aerial bombardment."[12] The US commander, General Petraeus, approved the deployment of a company of fourteen tanks.

As if to prove that the ghosts of Vietnam were still not fully exorcised, the American military remained sensitive to any media assertions of an "escalation" of the war; Lapan went on to explain that "These things happen all the time." There were also concerns that Afghan citizens might view tanks as too much like the Soviet occupation, so local commanders worked with local leaders and *shura* (a conference of local leaders) to minimize concerns.[13]

After leaving Iraq Robert Bodisch served in intelligence billets, including anti-drug operations along the Mexican border, and attended the Command and General Staff College at Fort Leavenworth. In early 2010 he was assigned to 2nd Tank as the XO, then sent to Afghanistan as a planner. Many Marines had been baffled by the refusal to deploy tanks, given their force-protection advantages: "When R[egional] C[ommand] South, the Army command, was seeing how successful the Marine tanks were, they did submit a request for tank forces. But by that time point ISAF had reached the US force-manning levels, the force cap. . . ." Plans were underway to fly in armor, but "In early 2011 the decision was made that the Army would not bring tanks in."

The decision was a disappointment, since "We were losing so many people, and these MRAPs and these other vehicles, although much more survivable than any other wheeled vehicle we had, it's just not the same as a tank."

Bodisch went on to agree with the Danish analysis: "Back in Iraq, we were the weapon of choice, in every aspect, because we were a precision-fire weapon. You were able to limit collateral damage, and you were able to provide pinpoint fires in exactly the areas the riflemen, the infantry, on the ground wanted. . . . In Afghanistan, in which collateral damage was probably even more sensitive than in Iraq, you don't have that capability."

After fourteen months Marc Tucker returned to Bravo, 1st Tank as the 3rd Platoon leader, then was moved to Scout Platoon and sent to Romania and Bulgaria to train troops for Afghanistan. Then in late 2010 Tucker found that the battalion would stand up a Delta Company under now-Major Dan Hughes, with him as XO.

Hughes had filled a number of billets including recruiting officer candidates, and a staff position in Afghanistan in 2009–2010. "Colonel Barett pulled me aside and said 'Hey Dan, they're gonna send a tank company to Afghanistan. It's gonna be your company. We got some work to do.'" Delta was a rump company that normally provided tanks for the MEU deployments, not a fully trained company. Marc Tucker was put in charge of training.

Hughes' primary concern, given the environment, was combat readiness of the trains. Counter-IED and counter-ambush exercises were conducted at Twentynine Palms, the first such to involve tanks. Additional emphasis was on training the support personnel in infantry skills.

Considerable horse-trading was involved in acquiring personnel, including experienced infantry to protect company trains. Tucker: "I know Dan won't tell you this, but Dan was loved, absolutely loved, by every single one of those Marines, because he gave them exactly what they wanted. They wanted tough, realistic, and hard training. They wanted to go down-range and fight for what was right. [What] Dan not only promised them, but gave them."

The biggest problem was defining the mission: "That was never very clear whatsoever. . . . If you can give me some kind of guidance, then we can go from there. We had very, very little so far as something that was hard. . . . As the tanks were being flown in to [Camp] Leatherneck . . . we were still taking care of loose ends." The advance personnel were flown in after Christmas, 2010.

Looking back, Tucker thought that LtCol Thomas J. Gordon had "inculcated his Marine officers with an idea that we need to be prepared. He did what I think great leaders do, which is they prepare you without you really know you're being prepared."

Always foremost in Hughes's mind was "We're the first tank company to go to Afghanistan. We need to do this right. We can't show up blowing everything up. We can't go there and not be able to be used. If we go there and we can't be used, and we're always broken down, and we never can help the infantry, well when's the infantry ever gonna want you?"

Upon arrival in Afghanistan in late 2010 Hughes was pleased to find that the infantry staff was receptive to suggestions about how the tanks might best be used. Tucker: "The mission was still kind of murky. . . ." although the battalion staff had considered possible roles. Tucker was surprised at how quickly the tactical potential for tanks was forgotten in light of Fallujah and other recent experiences. "Their initial impression was 'Well, we could use you guys as a QRF (Quick Reaction Force). . . .which was exactly what we *weren't* trying to do. That's very reactive, versus being very proactive." The tankers preferred to embed platoons where "We could provide overwatch, we could provide interdiction." When small sub-units are attached to larger units, "You run the risk of those members kind of being outcasts, being on the outer circle. With BLT 3/8 it was nothing of the sort." In an odd reversal of the situation that had caused so many problems for tanks in the past, Delta Company had brought its organic logistical support, while 3/8—which had been flown in from a ship—was short of vehicles and other necessities. "So we pushed all the logistic capability out to each one of his locations. . . ."

Hughes' company was attached to RCT-2, who placed one platoon in direct support of 2/4. Hughes: "The remaining ten tanks and the trains were going to operate in direct support of Three-Eight." The tanks were further parceled out to support the scattered 3/8 companies who were tasked to protect road crews improving and paving Route 611, a major north-south road plagued by mines, IEDs, and ambushes. CAAT teams and a 2nd LAR platoon (mounted in MRAPs) escorted convoys on the road. Tucker: "Anybody and everybody had some pretty significant and intimate meet-and-greets with IEDs. . . . It didn't matter if you were a logistics convoy, it didn't matter if you were route-clearance, it didn't matter if you were tanks. The whole idea was to blow people up. . . ."

The CO of 3/8, LtCol Farrell Sullivan, had been Tucker's platoon leader from his days in 3/8 in Kosovo, and the tank company CO, Dan Hughes, had also worked with Sullivan when he was a sergeant involved in the much-publicized rescue of USAF Capt. Scott O'Grady in Bosnia.

The various tank platoons were soon split among various operating areas, particularly protecting the road crews and with Kilo 3/8 to the left bank of the Helmand River, an avenue of approach for Taliban and foreign fighters from the north. Hughes noted that this was the first section to be in steady contact with the enemy. "The Headquarters tanks, we operated with the Scout Section to give the BLT commander an additional maneuver element."

Hughes assigned a mechanic as a loader in each section to do quick fixes, a practice that paid off when some of his tank sections were deployed in isolation for up to ninety days.

The imaging capabilities of the tanks provided long-range observation in the open terrain, and tanks worked in cooperation with scout-sniper teams and drones. The combo allowed the Marines to both covertly observe enemy movement and call in artillery or air assets far beyond the range of their guns. Tankers and scout-snipers could covertly observe routine activities in towns and villages. Hughes said that the Marines ". . . used the SIGINT to monitor the [enemy] chatter to see how they responded. Obviously they didn't like it. But what it showed them was we could take these tanks and move them anywhere we wanted to."

A cat-and-mouse game ensued as the insurgents tried to assess the new tanks. "After about a week of this, they finally figured out that these things can kind of see you anywhere. And they can kind of shoot a long ways." The Afghans proved far more canny fighters than the Iraqis, and only one incident of mindless bravado stood out in Hughes's mind: "We did have one that came out of a compound and tried to shoot at a tank from the back of a motorcycle with an AK-47. That didn't turn out so well."

The various infantry commanders "owned" their battle-space, and at first the tankers with experience in Iraq were taken aback at the rules of engagement. Fires had to be approved at higher echelons, with the inevitable delay in response time. Tucker said that in Iraq ". . . there really was no restraint other than that officer who was on the scene. I originally took it almost like a young lance-corporal that doesn't understand ROE. Like 'This is BS'." However, "It was one of the checks and balances that had to happen." Time and experience with the observation capacities and direct-fire accuracy of the tanks led to some relaxation, and "Very quickly we were given the ability and the flexibility that we needed to operate as effectively as possible."

IEDs were the insurgents' weapon of choice. Hughes: "One of the big things that I think was a psychological blow [to the enemy]. . . . When tanks did hit IEDs, what we always tried to do was fix the tank and drive it away." Damage usually involved only road wheels or suspension arms

On 6 March 2011 the Headquarters Section of two tanks was providing security for a logistical convoy and visit by the MEU commander to Kilo Company positions on the west side of the Helmand River. Tucker: "The terrain was such that there was one way in and one way out," ideal conditions for an ambush along a two or three kilometer stretch of dirt road. "They knew that, and we knew that. Just about every time we went up in there, somebody got blown up." Heading back out, "We knew, obviously, that we had been in this area long enough that there were some folks out there watching what was going on." Dan Hughes led the convoy on the return trip, along the road they had traversed about an hour earlier, with Tucker trailing.

They passed an Air Force Route Clearance Team, and assumed the road was safe. The convoy stopped, and Tucker leaned inside the tank to radio the MEU that they were leaving their battlespace.

Hughes: "We had just come around a corner, and I turned around and saw the front of his [Tucker's] tank come behind me and as soon as I turned around I heard this 'THOOM', and you could feel the blast. . . . A piece of his track had blown clear through the top of the fender." Hughes watched as three sets of road wheels were blown up to eighty meters away.

Tucker: "About a two-hundred-pound directional IED blew up. Unfortunately my left arm was exposed over my caliber fifty and took a piece of shrapnel, what we think was an end connector. It blew through my arm and took out about a two-and-a-half inch piece of my radial bone and my radial artery." The impact cauterized the artery ". . . so I didn't have any initial issues with the potential for bleed-out. I didn't even know I was hit for about the first five minutes."

Tucker's ignorance of his injury resulted from a flap of loose fabric that closed back over the wound. "We checked the crew, and everything appeared to be fine." Tucker oriented his gunner to search for possible attackers with machine guns or RPGs. "Soon thereafter I realized that I didn't feel so great. After making sure my Marines were taken care of, I did like a full-body sweep and kind of pulled back this old tin-can type tear in the top of my CVCs (coveralls) over my left forearm. I pulled it back, and what I saw kind of blew my mind. I'm looking at about a softball-sized hole through my forearm. I called over the net, called Dan and let him know what's going on."

Hughes backed his tank against the front of Tucker's, "And by that time they were lifting him out of the turret and had a tourniquet applied. You could see his arm was pretty much wide open." Hughes loaded Tucker onto the back of his turret and told his driver to head for the closest site for a helicopter landing zone at the top of a nearby hill.

"I was talking to him, asking him how he's doing, and he says 'How they gonna do a man like this? I'm not goin' out this way.'"

The problem was not so much fear of additional attack. "Our little convoy had no corpsman with us. I took one look at his arm and said, 'This guy needs a corpsman'."

Tucker: "The Brits flew a [CH] forty-seven in and evaced me out of the battlespace."

The planning for various contingencies paid off for Tucker. "I believe the total [evacuation] time line for me was thirty-seven minutes."

After a short ride, "Right before the lights went out with the anesthesia, the British doctor looked at me and told me that when I woke up I probably wouldn't be there. Definitely encouraging words. Lo and behold, I woke up and still had my arm." Not wanting to leave, "Between fighting and pleading with the [medical] staff there, to keep me in country, they pretty much laughed at me." Heavily sedated, Tucker was flown to Bagram and two days later to Landstuhl for several surgeries, then to Bethesda for a four and a half month stay and fourteen surgeries.

The small convoy had no recovery vehicle, and the recovery team came under attack. Hughes: "My Forward Air Controller, Captain Theissen, started engaging with the [M]-240, and one of my scouts, Staff Sergeant Echols, had jumped into the vehicle and started using the TC's fifty-cal." Additional tanks began to engage the insurgents from nearby high ground.

The tank, burdened by its additional belly armor to protect against mines and with missing wheels, proved difficult to budge up the steep grade.

One tank hooked a cable to the front, and another pushed with a towbar. "You had a lieutenant in the front pulling and a captain in the back. One thing you don't want running the show right now is a couple of officers. . . ." so two senior NCOs took over the effort. The tank was dragged to the top of the hill, and recovered the next day with an M-88 protected by a Force Recon detachment.

Activity lulled during the April poppy harvest, and BLT 3/4 began to replace 3/8. Most of the tanks were redeployed to operate out of Salaam Bazaar near the market town of Now Zad, to give 3/4 time to familiarize themselves with the situation. Hughes found that many of the junior enlisted men had little experience working with tanks, so he assigned experienced infantrymen from his Scout Section to act as liaisons. "When you needed a tank, he knew how to talk tank. He knew how to paint the picture that the tanks needed, so they could do what needed to be done for that infantry squad." Captain Theissen also went with foot patrols to coordinate air support. The tanks patrolled as a visible threat, which the insurgents tended to avoid.

Hughes thought that the Afghans had neither the experience nor the weapons to effectively fight tanks, and "The bite you got back wasn't worth it." The tankers worked out a tactic of sending a section out into the Operating Area and set up a surveillance position. "They would just move it, and as the patrols went out they were able to overwatch. It took a while, but eventually they're gonna shoot at somebody. They're gonna take their chances, and then we're able to affect them that way."

Tanks were also able to observe insurgents planting IEDs. "They were trying to drive away, but the tanks could go anywhere that motorcycle could go, so the tanks would stop them and bring up a mounted element to actually detain them."

The great lesson in Hughes' mind was that the tank's sensor systems were of paramount importance. "Even if we don't fire a single round for you, the far-target locator and our ability to acquire a target and acquire a grid [location] out to six thousand meters, that alone brings you a tremendous combat multiplier." Coupled with this was the tank's value as a sort of "forward deployed quick reaction force," whose firepower and mobility allowed it to be quickly pulled off one mission and assigned another.

At the bitter end of a long supply line, the tanks were sometimes not in tiptop condition, but were "still better than anything the infantry had." In some cases tanks were reduced to conditions where the engine and armament could not be operated simultaneously: the tank could either drive, or operate the turret with the auxiliary hydraulics. "Then when that went out, the gunner was hand-cranking it [to move the turret]."

In 1 August 2011 Hughes's company left Afghanistan as part of the normal force rotation, replaced by Captain Matthew Steiger's company from 2nd Tank.

By year's end the war had settled into a deadly routine as Marines repeatedly swept the countryside and stalked insurgents. As always, troops made their lives as comfortable as possible in the circumstances while trying to interdict the flow of bulk opium down the Helmand Valley. The "War Pigs" of C Company, 3rd LAR were typical: not only MREs, but luxuries like camp stoves, cooking pots, weight lifting equipment, and a stash of Cajun spices belonging to SSgt Garrette Guidry, C Company's maintenance chief, cluttered the vehicles. "I just hope all this doesn't spoil my love for camping," said company First Sergeant Justin Owens.[14]

The LAR units would patrol and provide armored security for the "defensive" civic action programs through mid-2012. Offensive actions were smarter versions of the old "search and destroy" sweeps of the Vietnam era. Emphasis was on intelligence gathering, observing the Taliban for months before tank-infantry sweeps like Operation JAWS in late-May early-June, 2012. The payoff was lopsided victories, in this case 50 insurgents killed, with no Marine casualties.[15]

There was a price to be paid for this success; unending maintenance. Personnel were rotated, but the units used the same tanks and LAVs. The flour-fine abrasive "moon dust" seeped into everything, and the jolting of the tank moving over the rugged terrain was brutal on the complex hydraulics and electronics. Mechanics like LCpl Lucas Walsh discovered that "Finding a [hydraulic] leak is like finding a needle in a haystack." Electronics were worse.[16]

The primary role for the tanks continued to be cooperation with scout-sniper teams. In September 2013 LtGen John Toolan noted that in a ten-day period, "The tank and sniper teams have contributed to about 50 enemy insurgents killed, using the snipers as sharpshooters and the tanks for the surveillance capability. It's really a great combo. . . ."

But like all Marines, the days of the tanks were numbered. The last tank unit to withdraw from Afghanistan was Delta, 2nd Tank Battalion in late 2013. The LAVs would soon follow.

NOTES

1 Grattan, Battalion Updates, June 1, 2009.
2 Anonymous, *Operation Khanjar restores government control in Khan Neshin*; Chandrasekaran, *Marines Meet Little Resistance in Afghan Push*.
3 Anonymous (Staff Report), *2 staff NCOs among latest Marine casualties;* Anonymous (Staff Report) *Lejeune NCOs die in Afghanistan*.
4 Curvin, *2nd LAR Marines interact with locals in southern Helmand*.
5 Grattan, *Message from the CO—August 31, News From Abroad*
6 Grattan,. *Update From Our Commanding Officer, September 11, 2009*
7 Anonymous, *Corpsman with 2nd LAR killed in Afghanistan*.
8 Reinhold, Christian, *Danish Tanks in Serious Fire-Fight in Afghanistan*
9 Cadieu, Trevor, *Canadian Armor in Afghanistan*, p. 10, 19–20.
10 GlobalSecurity.org, *Mine Resistant Ambush Protected (MRAP) Vehicle Program*
11 Estes, Kenneth, *Marines Under Armor*, p._____.
12 Col. Dave Lapan, cited in Garamone, Jim, "U.S. Tanks En Route to Southwestern Afghanistan"
13 Ibid.
14 Ross, Jeremy, *War Pig Homes: Cramming Comfort Into Life On The Road*
15 Lamothe, Dan, *Marines attack insurgents in Zamindawar area of Afghanistan*
16 Buckwalter, Brian, *Tanks Support Infantry Marines In Afghanistan*

Epilogue

We sleep safely at night because rough men stand
ready to visit violence on those who would harm us.
—attributed to George Orwell

A LTHOUGH THE LAST Marines are scheduled to exit Afghanistan in late 2014, this is an unfinished story. The pessimist would say that given human nature, it will forever remain unfinished.

In the closing days of the Iraq occupation Mike Mummey observed, "Now they've got to relearn how to do the big blue arrow fight. . . . But they're starting to relearn. When I was out there two weeks, the kid who was the colonel of the battalion—I was his first platoon sergeant—he's getting the battalion back into the big blue arrow mentality. They're going out and doing maneuver warfare again instead of convoys, or operating out of FOBs, stuff like that."

Yet the "big blue arrow" of mechanized maneuver warfare may again be withering. The mantra of "lightness" is ever recurrent in the Corps. As Dennis Beal points out, "Going back to a MEU size operation, that's our pedigree." Indeed, the "The MEU is the basis for our operational concepts. . . . They go in and do what they need to do, and then we follow in with either a brigade or the Army follows on, or whatever. That's not gonna change. That was the case before the ninety-one war, and before the . . . two thousand and two war."

The vision of itself as a light force designed to seize advanced bases, or to force an entry into enemy territory and leave the heavy land fighting to follow-on Army forces, is a cherished self-image that dates back to the Advanced Base Force concepts of the 1930s. The only problem is that it has almost never worked. Politicians and very high-level military planners alike have assured that the Marines have all too often been asked to assume the one role they have never wanted, acting as a sort of adjunct army. In fact, only twice in its institutional history has the Corps functioned in its "forced entry" role and then been relieved for prolonged operations by the US Army or allied forces: the Bougainville campaign of 1943–1944, and the ill-fated Operation RESTORE HOPE in Somalia, 1992–1994. An added complexity of the early twenty-first century is that even the most petty of dictators can—and for reasons of prestige and power must—maintain an armored force. The lesson of history is that the Corps, if it is to be a true force in readiness, must in turn maintain the capability of undertaking mechanized land operations.

The emphasis on "lightness" inevitably leads to reconsideration of the role of the tank. Robert Bodisch among many others is concerned about the seemingly cyclical decline in the understanding of the role of heavy armor in the Corps' primary mission. "The Commandant issued a guidance last year about lightening the MAGTF (Marine Air Ground Task Force). A lot of people that are in that process of figuring out ways of lightening the MAGTF, one of the easiest ways is to get rid of the tanks. . . . In the armor community perspective, that's not what we think the Commandant means. The Commandant means get rid of the bloated C-Two (Command and Control) structure that keeps growing and taking up valuable ship space at the expense of combat power."

Bodisch wrote a response pointing out that ". . . the Marine Corps is the only force in the word that has an expeditionary armor capability. Nobody else has armor forces pre-boated that can go into an amphibious environment." But that capability comes at a price in support personnel, logistical complexity, and simple deck space that MEU planners and commanders are always reluctant to accept. That is the overwhelming reason that the LAV has replaced the tank as the armored vehicle of choice in the MEU: five of the lighter vehicles can be transported and supported in the place of one tank.

So what is the future of armor—particularly heavy armor—for the Marine Corps?

Ever since the ill-fated development program for an "expeditionary tank" produced the Marmon-Herrington tankettes of the 1930s, the Corps has adopted and used Army-designed tanks, despite fundamental differences in operational requirements. The long-cherished idea of a true amphibious tank has repeatedly foundered on the fundamental fact that heavy armored vehicles just don't float very well.

In the short term improved versions of the M1 series are the only viable option. Dennis Beal believes that despite their age the M1 series tanks will soldier on for decades, at least twenty to thirty more years. "You pay a lot of money for these things. . . . It's not a car you drive for three years, and you put it up and buy another one. These things are five million bucks apiece. . . . You upgrade 'em, you mod[ify] 'em, . . . you product-improve 'em, you do all that. But what about the next generation?"

A modern complication is that technological complexity, and its attendant soaring costs, have doomed recent attempts by the chronically fiscally-strapped Marine Corps to acquire new mission-fundamental weapons systems, even for well-established mission requirements. One recent victim was the Advanced Amphibious Assault Vehicle (AAAV), a complex house-sized "swimming Bradley." Armed with a 30mm chain gun, the AAAV was to be a sort of military Transformer,™ capable of changing its hull shape to carry a squad of Marines at high speed from ships far out at sea and yet stand some chance of survival in land combat.

In the end the AAAV program was victim to the seemingly inexorable feedback loop of constantly evolving design and soaring costs. There has been no recent national emergency that would provide motivation to short-circuit this escalating spiral to provide a weapons system that, like the American tanks of World War II, would not necessarily be the technological best but affordable and adequate for the mission.

The conundrum of a future expeditionary tank that balances the three tank fundamentals of armored protection, firepower, and mobility is for the Marine Corps far more complex. Since the mid-twentieth century the main battle tank has in most armies evolved into a machine dedicated to killing its own kind. For the Marine Corps the expeditionary tank will likely remain wedded to its dual role of not only killing enemy tanks, but providing precision direct-fire support for the infantry. As retired Army Lieutenant Colonel and military theorist Ralph Peters has pointed out, the Corps will likely find itself increasingly embroiled in urban combat, the type of battle for which the main battle tank may be less well-suited, but for which some sort of protected direct support weapons system, i.e., a specialized future generation tank, will be essential.[1]

Technological advances make it increasingly likely that the generation-after-next main battle tank will further evolve into something out of a science fiction novel, powered by as-yet unproven propulsion technologies, and armed with a futuristic weapon such as a hypervelocity rail gun that kills at fantastic ranges by pure kinetic energy. Of course the problem for the Marine Corps, with its requirement for infantry support, is that the expeditionary tank will still require a weapon capable of reducing hardened enemy defenses—that is, one capable of firing a chemical explosive/fragmentation round. This duality of mission may at last force the development of the long-dreamt of specialized expeditionary tank.

For the Marine Corps of the future the tank is not dead. But it surely will be nearly impossible to visualize.

NOTES
1 Peters, *Our Soldiers, Their Cities*.

Where Are They Now?

CHRIS AKRIDGE completed his enlistment, returned to college for his last semester, and became a high-school teacher for four years. He returned to graduate school and in 2012 earned a Master's Degree in social work, with the intention to work with veterans' services.

DENNIS BEAL worked in development of future armor for the Marine Corps before his retirement. He is now a program administrator at Texas A&M University.

MAJOR ROBERT J. BODISCH is still on active duty.

MASTER SERGEANT SAMUEL CRABTREE served as provisional company XO, and served an additional deployment to Romania. He found a direct application of his military skills as a heavy mobile equipment mechanic, and is currently a crew leader for the US government, refurbishing tanks. "I actually have to get dirty and work, where in the Reserves I'm not out on the tanks."

BUSTER DIGGS retired in 1995. "Once you make Colonel, that's it in the Marine Corps in tanks." Like a surprising number of Marines, Diggs became a middle-school teacher, and is now a substitute teacher.

TIM FRANK attended college, and became a Marine aviator, flying F-18s in OIF-1.

JAMES GONSALVES left the Corps after ten years, but applied for reinstatement after 9/11. He served two tours in Iraq, in a composite engineer group and as a Civil Affairs team leader supporting LAR battalions at Korean Village. He is in the IT industry.

BILL HAYES regretted not seeing more action in OIF-1, but after Fallujah "It was 'Okay, I rated the Combat Action ribbon. Let's kind of EAS out of the Marines, and just be glad I didn't die.'" He attended college, works for the State Department, and is in graduate school.

DAN HUGHES was assigned to Fort Benning, Georgia as an instructor in the Captain's Career Course, teaching tactics and leadership.

KEVIN KESSINGER went on recruiting duty, then left the Marines after thirteen years and entered graduate school, eventually becoming a mechanical engineer in specialty plastics molding and sales. "I didn't really miss the Marine Corps, but I missed the people. I missed the *quality* of people, really."

RICK MANCINI served four tours in tanks ("a real anomaly") including the command of 1st Tank, at Headquarters Marine Corps, and retired as a colonel. He worked as a security consultant in Afghanistan, and now "semi-retired," runs his own security consulting company.

MICHAEL MARTINEZ became a Master Gunnery Sergeant (E-9), the highest technical enlisted rank in the Marine Corps. He is with the US Marshall Service and is now a Reserve instructor in the Marine Corps Tactics and Operations Group. He occasionally works in Afghanistan under the auspices of the State Department.

MIKE MUMMEY retired in June 2004. He works in the construction industry, in sales, scheduling, and materials acquisition.

KEVIN MORONI was injured in a low-speed motorcycle accident after leaving the Corps, with resulting closed-head trauma.

JOHN POLIDORO was promoted to Lt Colonel; he is still on active duty.

MARC TUCKER recovered about eighty per cent use of his arm: "This is probably about as good as it's gonna get." He went back to the Officer's Basic School yet again as an instructor, "to take all the experiences I had in the Marine Corps and give them back to people who were going to go out and *do*."

References Cited

NOTE: Where information is derived from a US Government or private website, the full web address is provided. However websites are ephemeral; content may have changed, or the website may have disappeared.

Ackerman, Elliott L., *Relearning Storm Troop Tactics: The Battle for Fallujah*, Marine Corps Gazette, vol. 90, no. 9. p. 47–54, 2006.

Adkins, Mark, *Urgent Fury–The Battle For Grenada*, Lexington MA, Lexington Books, 1989.

American Forces Press Service, *'Iron Fist' Offensive Pummels Terrorists In Iraq*, http://www.defenselink.mil/news/newsarticle.aspx?id=18174

Anonymous (Staff Report), *2 staff NCOs among latest Marine casualties*, http://www.marinecorpstimes.com/news/2009/07/marine_afghan_casualties_071509w/#

_____, *5th MEB Deployment to SWA*, USMC summary report, unpaginated, undated.

_____, *A War of Logistics* (Interview with BGen Charles C. Krulak) in Melson, Charles D, Evelyn A. Englander and David A. Dawson, *U. S. Marines in the Persian Gulf, 1990–1991: Anthology and Annotated Bibliography*, Washington DC, History and Museums Division, 1992. Reprinted from U.S. Naval Institute Proceedings, November 1991.

_____, *Battle of al-Qaim*, http://en.wikipedia.org/wiki/Battle_of_Al_Qaim

_____, *Corpsman Up!*, http://www.i-mef.usmc.mil/div/3lar/EventPhotos/CorpsmanUp.asp

_____ (Staff Report), *Corpsman with 2nd LAR killed in Afghanistan*, Marine Corps News Room, http://www.marine-corps-news.com/2009/09/corpsman_with_2nd_lar_killed_i.htm

_____ (Staff Report) *Lejeune NCOs die in Afghanistan.* http://www.marinecorpstimes.com/news/2009/07/marine_afghan_kias_072409w/?FORM=ZZNR9

_____, *Operation Iraqi Freedom Iraq—1st Tank Battalion, After Action Report, (Narrative)*, 50p.

267

_____, *Operation Iron Fist—Operation Kabda Bil Hadid*, http://www.globalsecurity. org/military/ops/oif-iron-fist_2005.htm

_____, *Operation Khanjar restores government control in Khan Neshi*, CENTCOM Press Release http://www.centcom.mil/en/press-releases/operation-khanjar-restores-government-control-in-khan-neshin.html

_____, *Rolling With the 2d Marine Division* (Interview with Lt Gen William M. Keys) in Melson, Charles D, Evelyn A. Englander and David A. Dawson, *U. S. Marines in the Persian Gulf, 1990–1991: Anthology and Annotated Bibliography*, Washington DC, History and Museums Division, 1992. Reprinted from U.S. Naval Institute Proceedings, November 1991.

_____, *Special Trust and Confidence Among the Trail Breakers* (Interview with Lt Gen Walter E. Boomer) in Melson, Charles D, Evelyn A. Englander and David A. Dawson, *U. S. Marines in the Persian Gulf, 1990–1991: Anthology and Annotated Bibliography*, Washington DC, History and Museums Division, 1992. Reprinted from U.S. Naval Institute Proceedings, November 1991.

_____, *SSgt Donald C. May Jr.*, http://donaldcmayjr.memory-of.com/About.aspx

_____, *The 1st Marine Division in the Attack*, (Interview with MGen J. M. Myatt) in Melson, Charles D, Evelyn A. Englander and David A. Dawson, *U. S. Marines in the Persian Gulf, 1990–1991: Anthology and Annotated Bibliography*, Washington DC, History and Museums Division, 1992. Reprinted from U.S. Naval Institute Proceedings, November 1991.

_____, *The Great Game*, http://en.wikipedia.org/wiki/The_Great_Game

_____, *This Was No Drill* (Interview with Major General John I Hopkins), in Melson, Charles D, Evelyn A. Englander and David A. Dawson, *U. S. Marines in the Persian Gulf, 1990–1991: Anthology and Annotated Bibliography*, Washington DC, History and Museums Division, 1992. Reprinted from U.S. Naval Institute Proceedings, November 1991.

Applewhite, Raymond L., and Eric Schwab, untitled Navy news release NNS0408 12-05http://www.marinecorpsmars.com/Honors%20Pages/References/luis_e_fonseca.htm

Ballard, John R., *Fighting for Fallujah*, Westport CT, Praeger Security International, 2006.

Battlefield Assessment Team, Armor/Antiarmor Team, *Armor/Antiarmor Operations in South-West Asia*, Quantico VA, Marine Corps Research Center, Research Paper #92-002, 1991.

Belvoir Research, Development and Engineering Center, *Desert Storm Countermine Equipment*, Fort Belvoir VA, brochure, undated.

Bennett, Drew A., *Minefield Breaching: Doing the Job Right*, Armor Magazine, vol. I, no.4, July-August 1992, p. 19–21.

Bonadonna, Reed R., *Interview with Major William Peeples and First Sergeant Roger Huddleston*, CD13266, Oral History Collection, Quantico VA, Alfred M. Gray Research Center, 2003.

Buckwalter, Brian, *Tanks Support Infantry Marines In Afghanistan*, http://marines.dod live.mil/2012/08/20/tanks-support-infantry-marines-in-afghanistan/

Cadieu, Trevor, *Canadian Tanks in Afghanistan*, Canadian Army Journal vol10.4, p. 5–25, 2008.

Chandrasekaran, Rajiv, *Marines Meet Little Resistance in Afghan Push*, Washington Post, July 3, 2009.

Chang, Tao-Hung, *The Battle of Fallujah: Lessons Learned on Military Operations on Urbanized Terrain (MOUT) in the 21st Century*, Naval Science Thesis, http://sa.rochester.edu/jur/issues/fall2007/chang.pdf, 2008

Cloud, David, *The Fourth Star: Four Generals and the Epic Struggle for the Future of the US Army*, New York, Crown Publishing, 2009.

Copley News Service, *State of the Art in Tank Warfare*, Houston Post, February 3, 1991, p. A-17.

Corwin, Tony L., *BLT 2/8 Moves South*, in Melson, Charles D, Evelyn A. Englander and David A. Dawson, *U. S. Marines in the Persian Gulf, 1990–1991: Anthology and Annotated Bibliography*, Washington DC, History and Museums Division, 1992.

Cureton, Charles C., *Interview with Captain Roger L. Pollard*, CD10441, Oral History Collection, Quantico VA, Alfred M. Grey Research Center, 1991. (Note that there is an additional, unidentified speaker in the interview, in fact the company XO).

_____, *Interview with Lieutenant David Kendall*, CD10441, Oral History Collection, Quantico VA, Alfred M. Grey Research Center, 1991.

_____, *Interview with members of OCD attached to Combat Engineer Detachment, Task Force Papa Bear*, CD10441, Oral History Collection, Quantico VA, Alfred M. Grey Research Center, 1991.

Curvin, Michael, *2nd LAR Marines interact with locals in southern Helmand*, http://www.iimefpublic.usmc.mil/public/InfolineMarines.nsf/ArticlesListingRead-Current/A8858C734BD583A28525761E004D4CE8?OpenDocument.

Dacus, Jeffrey R., *Bravo Company Goes to War*, Armor Magazine, vol. C, no.5, September–October 1991, p. 9–15.

Danish Military Press Release, *Afghanistan: Alle danske soldater er nu tilbage i deres respektive lejre i Helmand* provinsen, 25 July 2008, http://forsvaret.dk/HOK/Nyt%20og%20Presse/ISAF/Pages/Statusoverfredagensbegivenheder.aspxx

Dunfee, David R., *Ambush Alley Revisited*, Marine Corps Gazette, vol. 88, no. 3, p. 44–46.

Estes, Kenneth W., Colonel Dennis W. Beal briefing notes, 14 April 2003.

_____, Interview with Dennis Beal, 18 March 2009.

_____, Interview with Major Mike Campbell, undated.

_____, "Learning Lessons From The Gulf?" *Marine Corps Gazette*, November 1998, p. 92-93.

_____, *Marines Under Armor*, Annapolis MD, Naval Institute Press, 2000.

_____, *US Marine Corps Operations in Iraq, 2003–2006*, Quantico VA, US Marine Corps History Division, 2009. Available online: http://www.fas.org/irp/doddir/usmc/iraq03-06.pdf. Note that pages cited refer to the pre-publication manuscript.

Fahy, Richard, *Interview with Major Tim Frank*, CD13810, Oral History Collection, Quantico VA, Alfred M. Gray Research Center, 2003.

Fike, Jenna, *Interview with Major Jason Gaddy*, Operational Leadership Experiences Project, Combat Studies Institute, Ft. Leavenworth, KS, 2009.

_____, *Interview with MAJ Robert Bodisch*, Operational Leadership Experiences Project, Combat Studies Institute, Ft. Leavenworth, KS, 2009.

Folsom, Seth W. B., *The Highway War*, Washington DC, Potomac Books, 2006.

Foulk, Vincent L., *The Battle for Fallujah*, Jefferson NC, McFarland, 2007.

Freitus, Joe and Chris Freitus, *Dial 911 Marines—Adventures of a Tank Company in Desert Shield and Desert Storm*, McCarran NV, New American Publishing, , 2002.

Galuzska, Doug, *Interview with MAJ Daniel Benz*, Operational Leadership Experiences Project, Combat Studies Institute, Ft. Leavenworth, KS, 2005.

Garamone, Jim, "U.S. Tanks En Route to Southwestern Afghanistan," *Washington Post,* November 19, 2010.

Gilbert, Ed and Allen Swan, *T-72 in Iraqi Service,* Katy TX, Full Detail, not paginated.

GlobalSecurity.org, *Mine Resistant Ambush Protected (MRAP) Vehicle Program,* http://www.globalsecurity.org/military/systems/ground/mrap.htm

Glover, H., *Command Chronology Report for Calendar Year 2003, Company A, 8th Tank Battalion,* 2004.

Goodwin, Jim, *Marine tank unit crucial to stabilization of Iraq,* http://www.globalsecurity.org/military/library/news/2005/01/mil-050127-usmc01.htm

Gott, Kendall D., *Breaking the Mold: Tanks in the Cities,* Fort Leavenworth KS, Combat Studies Institute Press, 2006.

Grattan, Thomas E., *Battalion Updates, June 1, 2009* http://www.iimefpublic.usmc.mil/public/InfolineMarines.nsf/(ArticlesRead)/6FCCB959872D11AD852575CB00469468

_____, *Message from the CO—August 31, News From Abroad,* http://www.iimefpub-

lic.usmc.mil/public/InfolineMarines.nsf/ArticlesListingReadCurrent/FB70321
472A597E18525762C007F0063?OpenDocument

_____, *Update From Our Commanding Officer, September 11, 2009 Message From the
CO* http://www.iimefpublic.usmc.mil/public/InfolineMarines.nsf/(Articles
Read)/6288E34E56C9B05E8525762E005A96CE

Harris, Carroll C., *Interview with Second Lieutenant Aaron Klein*, CD13810, Oral History Collection, Quantico VA, Alfred M. Gray Resrach Center, 2003.

Hopkirk, Peter, *The Great Game: Struggle for Empire in Central Asia*. New York, Ko-
dansha International, 1992.

Ives, Christopher, *Interview With Major William W. Johnson*, Operational Leadership
Experiences Project, Combat Studies Institute, Ft. Leavenworth, KS, 2005.

Johnson, David E., Adam Grissom, and Olga Oliker, *In the Middle of the Fight*, Santa
Monica CA, Rand Corporation, 2008.

Jones, Charles, *Boys of '67*, Mechanicsburg PA, Stackpole Books, 2006.

Knowlton, *Troops Seize Airstrip Near Taliban Base : U.S. Marines Land in Afghanistan*,
New York Times, November 27, 2001.

Kozlowski, Francis X., *The Battle for An-Najaf*, Washington DC, U.S. Marine Corps
History Div., 2009. Note that although this is a secondary source, primary
source materials are extensively documented.

Kreisher, Otto, *Persian Gulf War: Marines' Minefield Assault*, The Quarterly Journal
of Military History. Summer 2002.

Kyle, R. Logan, *Wolfpack Captain and Iraqi Interpreter Reunite at Fort Irwin*, Twentyine
Palms CA, Observation Post (Marine Corps Air Ground Combat Center) v.
52, issue 5.

Lamothe, Dan, *Last U.S. Marine Tank Unit Headed Home From Afghanistan*, http://
blogs.militarytimes.com/battle-rattle/2013/09/05/last-u-s-marine-tank-unit-
heading-home-from-afghanistan/

_____, *Marines attack insurgents in Zamindawar area of Afghanistan*, http://blogs.military-
times.com/battle-rattle/2012/06/13/marines-attack-insurgents-in-zamindawar-area-of-
afghanistan/

Landers, Jim, *Ambush Costly for Marine Battalion*, Dallas Morning News, April 9,
2003.

_____, *The Marines' 2nd Tank Battalion Used Speed and Armor to Make Quick Work of
Saddam Hussein's Regime*, Dallas Morning News, May 18, 2003.

Lessard, Lawrence, *Interview with Major Dave Banning*, Operational Leadership Experiences Project, Combat Studies Institute, Ft. Leavenworth, KS, 2006.

_____, *Interview with MAJ John Polidoro, Part I*, Operational Leadership Experiences
Project, Combat Studies Institute, Ft. Leavenworth, KS, 2007.

_____, *Interview with MAJ John Polidoro, Part II*, Operational Leadership Experiences

Project, Combat Studies Institute, Ft. Leavenworth, KS, 2007.

Livingston, Gary, *An Nasiriyah*, expanded 2nd edition, North Topsail Beach NC, Caisson Press, 2003.

Lohse, Eckart, *Leopardenjagd am Hindukusch*, 16 March 2008, Frankfurter Allgemeine Zeitung, 16 March 2008, p. 6., http://www.faz.net/s/RubDDBDABB945 7A437BAA85A49C26FB23A0/Doc~E74D84BBBF12B478BBD4C7901BB39 BF59~ATpl~Ecommon~Scontent.html

Lowry, Richard S., *Marines in the Garden of Eden*, New York, Berkley Caliber, 2006. See also http://www.militaryhistoryonline.com/iraqifreedom/articles/gardenofeden.aspx

_____, *The Gulf War Chronicles*, iUniverse, 2003.

Michaels, G. J., *Tip of the Spear*, Annapolis MD, Naval Institute Press, 1998.

Millett, Allan R., *Semper Fidelis: The History of the Unites States Marine Corps*, New York, MacMillan, 1980.

Moore, Molly, *Allies Used a Variation of Trojan Horse Ploy,* in Melson, Charles D, Evelyn A. Englander and David A. Dawson, *U. S. Marines in the Persian Gulf, 1990–1991: Anthology and Annotated Bibliography*, Washington DC, History and Museums Division, 1992. Reprinted from The Washington Post, 17 March 1991.

_____, *Out Front At the Front: Marines Brace for Task of Clearing Mines,* in Melson, Charles D, Evelyn A. Englander and David A. Dawson, *U. S. Marines in the Persian Gulf, 1990–1991: Anthology and Annotated Bibliography*, Washington DC, History and Museums Division, 1992. Reprinted from The Washington Post, 19 February 1991.

_____, *Storming the Desert with the Generals,* in Melson, Charles D, Evelyn A. Englander and David A. Dawson, *U. S. Marines in the Persian Gulf, 1990–1991: Anthology and Annotated Bibliography*, Washington DC, History and Museums Division, 1992. Reprinted from The Washington Post, 14 April 1991.

Mortenson, Greg, and David O. Relin, *Three Cups of Tea*, New York, Penguin Books, 2006.

Mroczkowski, Dennis P., Restoring Hope: In Somalia with the Unified Task Force 1992–1993, online version, http://www.usmc.mil/news/publications/Documents/Restoring%20Hope%20In%20Somalia%20with%20the%20Unified%20Task%20Force%201992-1993%20PCN%2019000413500_2.pdf

Multi-National Corps Iraq Public Affairs, *Wolfpack Sees Operation Iraqi Freedom From Start to Finish,* http://www.i-mef.usmc.mil/articles/story206.asp

Peters, Ralph, *Our Soldiers, Their Cities*, Parameters (Magazine), Spring 1996, p. 43–50; available online at http://strategicstudiesinstitute.army.mil/pubs/parameters/articles/96spring/peters.htm

Pollard, Roger L. *The Battle for OP-4: Start of the Ground War,* Marine Corps Gazette, vol. 87, no. 7, p. 48–49.

Popaditch, Nick, and Mike Steere, *Once A Marine*, El Dorado Hills CA, Savas Beatie, 2008.

Pritchard, Tim, *Ambush Alley–The Most Extraordinary Battle of the Iraq War*, Paperback Edition, New York, Ballantine, 2007.

Reinhold, Christian, *Danish Tanks in Serious Fire-Fight in Afghanistan* (English translation of Danish military press release), http://www.casr.ca/ft-leopard-2a5-denmark-2.htm

_____, *Basrah, Baghdad, and Beyond*, Annapolis MD, Naval Institute Press, 2003.

Reynolds, Nicholas E., *Operation Just Cause: Marine Operations in Panama, 1988–1990*, Washington DC, History and Museums Division, U S Marine Corps, 1996.

Ringler, Jack. K., *U.S. Marine Corps operations in the Dominican Republic, April-June 1965*, Washington DC, Historical Division, U S Marine Corps, 1992.

Roggio, Bill, *Operation Iron Fist Near Qaim* http://www.longwarjournal.org/archives/2005/10/operation_iron.php#ixzz0l3vIXeNk&c

Ross, Jeremy, *War Pig Homes: Cramming Comfort Into Life On The Road*, http://marinesmagazine.dodlive.mil/2011/11/21/war-pig-homes-cramming-comfort-into-life-on-the-road/

Sachtlebem, James L., *Artillery Raids in Southwestern Kuwait*, in Melson, Charles D, Evelyn A. Englander and David A. Dawson, *U. S. Marines in the Persian Gulf, 1990–1991: Anthology and Annotated Bibliography*, Washington DC, History and Museums Division, 1992. Reprinted from Field Artillery, October 1991.

Sattler, John F., and Daniel H. Wilson, *Operation AL-FAJR: The Battle of Fallujah—Part II*, Marine Corps Gazette, vol. 89, no. 7, p. 12–24, 2005.

Simmons, Edwin H., *Getting Marines to the Gulf* , in Melson, Charles D, Evelyn A. Englander and David A. Dawson, *U. S. Marines in the Persian Gulf, 1990–1991: Anthology and Annotated Bibliography*, Washington DC, History and Museums Division, 1992.

Skaggs, Michael D., *Tank-Infantry Integration*, Marine Corps Gazette, vol. 89, no. 6, p. 41–43, 2005.

Smith, Elliott Blair, "Baghdad and Beyond," *USA Today* (online), http://www.usatoday.com/news/world/iraq/2004-04-08-baghdad-beyond_x.htm, 2004

_____, "Into Iraq," *USA Today* (online), http://www.usatoday.com/news/world/iraq/2004-04-07-baghdad-intoiraq_x.htm, 2004

Soussi, Alasdair, *50 Years Later, U.S. Marines Remember The 1958 U.S. "Intervention" in Lebanon,* Washington DC, Washington Report on Middle East Affairs, July 2008, pages 28–29.

Steinkopff, Mark, *Five Marines Honored For Service in Iraq War*, Jacksonville NC Daily News, March 23, 2004; also available at http://www.leatherneck.com/forums/showthread.php?t=13428

Storer, Ronald D., *Tank-Infantry Team in the Urban Environment*, Marine Corps Gazette, vol. 86, no. 5, p. 61–62, 2002.

Swift, Christopher, *Chris Swift in Desert Storm*, http://yankeetirade.com/Chris_Swift_Desert_Storm.html

Symonds, Craig L. and William J. Clipson, *Naval Institute Historical Atlas of the U S Navy*, Annapolis MD, Naval Institute Press, 2001.

Torres, Paul M., *Tank reservist brings police tactics to Anbar*, http://www.marines.mil/units/marforpac/imef/1stmardiv/Pages/MAR2.aspx

Turner, *Tanks with the MEU: A Team for Success*, Marine Corps Gazette, vol. 80, no.11, p. 39–41.

Ullman, Harlan K. and James P. Wade, *Shock and Awe: Achieving Rapid Dominance*, Washington DC, National Defense University, 1966.

Valliere, Samuel Baird, and Kristen A. Lasicam, *CSSB-1 helms mission to unearth sunken tank in northern Fallujah*, Leatherneck Magazine, http://www.leatherneck.com/forums/archive/index.php/t-13962.html, 2004.

Watson, Ben T., e-mail *Ramblings part II*, 14 April 2003, provided by Dennis Beal via Ken Estes

West, Bing and Ray L. Smith, *The March Up: Taking Baghdad With The 1st Marine Division*, New York, Bantam Dell, 2003.

Wikipedia, *History of Somalia*, http://en.wikipedia.org/wiki/History_of_Somalia

Winicki, Anthony A., *The Marine Combined Arms Raid*, Marine Corps Gazette, vol. 15, no. 11, p. 54–55, December 1991.

Wintermeyer, Paul W., *The Battle of al-Khafji*, Washington DC, U.S. Marine Corps Museums and History Div., 2008.

Youngquist, Sherry, "Pulled Through by His Buddies," *Winston-Salem Journal*, September 26, 2009; available online at http://www2.journalnow.com/content/2005/nov/06/pulled-through-his-buddies/

Zumwalt, J. G., "Tanks! Tanks! Direct Front!," *Naval Inst. Proceedings*, vol. 118, no. 7/1023, July 1992, p. 72–80.

Index